CLIMATE CHANGE AND ENVIRONMENT

HOW IT IMPACTS US ALL...

DR. VASUDEVAN RAJARAM,
KEITH OLSON & LYNN TIEDE

notionpress.com
INDIA • SINGAPORE • MALAYSIA

ISBN
Paperback 979-8-88667-570-2
Hardcase 979-8-90560-924-4

Contents

Testimonials

This book provides a broad spectrum of facts and figures related to climate change and its devastating impacts involving flood, fire and related disasters. It vividly covers the complex changes occurring in the environment due to global warming caused by greenhouse gas emissions. *Subijoy Dutta, Rivers of the World Foundation, USA.*

"The writers let you think about how you can deal with climate change. It unfolds with examples how you determine the root cause and find a way to address that." *Arun Seetharam, Advisor, Higher education, Govt of Karnataka.*

"This is not an alarmist voice. The authors present a quiet, dispassionate listing. The book asks readers to feel the gravity of the situation, and gently guides the way to serious action. It is an eye-opening experience, and it really goes to show that no one is too small to make a difference." Hari Haran Chandra, *AltTech Foundation, India.*

"Rajaram, Keith and Lynn break down the motivations and mindsets behind resistance to climate action with clarity and conviction." *Mary Conley Eggert, Founder, Global Water Works, Illinois, The USA.*

"The book shows us the need to fully understand how our lifestyles are impacting the planet. There's plenty of material out here for you to read, absorb, and share with students, peers and professionals." *Raghuveer, Green Accredited Professional, IGBC AP.*

"If you're looking for suggestions from a trustworthy source on how to combat climate change with personal actions then look no further." *Prof Kiran Nadgouda, Dean for Interns, College of IT in Karnataka.*

"The book demonstrates just how worthwhile and easy it is to sympathise, relate and work together to find solutions to better protect our environment." *Niranjan, Advisor, Green Jobs Skills Council, Govt of India.*

"Both the high end researcher and a graduate can relate to the actions the authors suggest." *Vasumathi Shetty, Education Consultant.*

"The book acts as a reminder that tackling climate change is very much about moving away from our patterns and finding simple ways to function as a local and yet global network. A must read for young professionals – be they business managers or engineers." *K L Mohan Rao Former Chairman, Builders Association of India.*

"The book's thoughts on environmental impact opens up a window for readers to think about human nature, and how we process the complex idea of climate change." *John Dalton, Faculty, University in Melbourne, Australia.*

"It tells you how much you can really do to make a difference. It shows you with many examples how reducing your personal carbon footprint is important and possible. So much so that PJMT has extended heartily its support to the Book. " *Payal Jain, Founder, Prem Jain Memorial Trust, New Delhi.*

"That the book is based on hard research and rich experience is evident. Their talent for creating numerous examples makes it an easy and compelling read. Through the examples they set, they help every learner cope with environmental changes with little support." *Prateek Bhaumik, Architect, Hyderabad.*

"In a lucid style, this book holistically introduces the existential threat of climate change and its multi-faceted impact on environment, society, and economy. This publication comes at the most suitable time as many countries around the world, including India, commits to Net Zero emission targets. I hope this inspires us, especially the younger generation, to go green in all our actions and ensure that Mother Earth remains a Sustainable Planet." *M Anand, Principal Counsellor, Indian Green Building Council (IGBC) and Vice Chair, WorldGBC's Asia Pacific Regional Network.*

"This book shows you how you can participate in the process of climate, if you're not part of a governmental or intergovernmental complex process. It guides the reader's voice to share, to walk the talk, to practice." *Usha Kini, First Founder, WOW Action Forum, India.*

"Environment and Climate Change is a book you'd like to keep around … to remind you that there is a solution, and that you can be part of it." *Sanjay Gupta, Bevels & Gears, Hyderabad.*

Also by Vasudevan Rajaram

Sustainable Mining Practices.
Solid and Liquid Management:...Waste to Wealth.
Ecology and Environment
Golden Giving: Everything You need to Know for an Enriched Socially Conscious Retirement.
From Landfill gas to energy: Technologies and Challenges.

The Authors

Dr. Vasudevan 'Raj' Rajaram is an environmental engineer with 44 years of experience in water and waste management projects in the United States and India. He founded Tetra Tech India in 1997 to transfer environmental technologies from the United States to India, and worked on the remediation of landfills, introduced Clean Technologies in many industrial sectors in India, and completed several large wastewater treatment projects. He has written four books on these subjects and these are: Sustainable Mining Practices – A Global Perspective; From Landfill Gas to Energy – Technologies and Challenges; both published by Taylor and Francis, United Kingdom, Liquid and Solid Waste Management – Waste to Wealth, published by Prentice Hall India; and Ecology and Environment, published by TERI Press. He has written a book entitled Golden Giving, published by Kindle Publishing. He has presented over 50 papers at several international conferences on the subjects of Mining, Water and Waste Management.

Keith Olson has worked, lived and studied in many locations through the years, the majority in mid-U.S.A. near Chicago, centred between the Atlantic and Pacific Oceans, where the continental climate comes with wide swings of temperature and precipitation. He also has lived in other countries and travelled extensively to all continents except Antarctica.

During his travels and decades teaching science and math, he developed a world view of respecting the environment the world around, including the necessity of addressing climate change. He has been and continues to be a member of several environmental organizations including one that specifically addresses the impact of climate on our lives. One significant activity of his is preserving a natural grassland, a sink for atmospheric carbon. He also lives a conservative lifestyle, therein reducing his impact on the environment. He usually gets around by walking and using public transportation. He and his wife have a small family and maintain a small environmental foot print, avoiding waste, recycling and keeping low demand on utilities. However, no person can avoid environmental impact. Even heating water for a cup of tea takes energy which, unless from green energy, will produce some greenhouse gases, thereby contributing to climate change.

Lynn Tiede is a public school teacher in New York City, NY with 29 years of teaching and curriculum writing experience, community activism, and non-profit work. She received her B.A. in Anthropology from Rhodes College in Memphis, TN and M.A. in Environmental Education from the City College of New York. After having also worked as Education Coordinator for the New York State Audubon Society, Lynn was chosen as a recipient of the Indo-American Fulbright program in Environmental Leadership in 2005. She spent several months studying ways that individuals, businesses, and the government of India were promoting and practicing habits of environmental sustainability. Currently at Columbia Secondary School for Math, Science, and Engineering High School, she teaches global history, participation in government, & economics, and as the school's sustainability coordinator works on a day to day basis to foster habits of sustainability with her students and colleagues. She is a member of two bodies, the Climate Education Leadership Team and the Climate and Resilience Education Task Force that are working to enhance education about climate change and sustainability in New York City.

Acknowledgements

- Dr. Rajaram dedicates this book to his wife who was patient while he worked on this book for 3 years. He is grateful to his parents for instilling a love of humanity and life-long learning. Several people helped in this book. Ms. Fatima Tuz Zebra contributed Chapter 5, and Ms. Annapurna contributed Chapter 6. Manjunath helped with all the graphics. Anand helped with cover design. Finally, Dr. Hari Haran Chandra of AltTech Foundation, Bengaluru, helped with publication of this book. His tireless effort to promote energy, built environment and water sustainability and help with the reaching of the book to a wide cross-section of readers, young and old, is gratefully acknowledged. Dr Agami Reddy for reviewing the chapter on energy, and Poorva Malik for help in the graphics that leaven these pages.
- Mr. Keith Olson acknowledges his co-authors for their versatile interests and patience. He also notes the many places, people and books whose lessons are too plentiful to list without being itself a book. And of course thanks his family for following him and leading him to locations, physical and of the mind.
- Lynn Tiede would like to acknowledge her students and all the activists and people across the world who are working to address climate change and all of the other environmental issues facing the world. My students in New York City and all of the wonderful people met during my travels and research in India inspire me daily to keep working for a greener world and solutions to climate change. Lynn would also like to acknowledge her family's support of her work and studies her entire life. Her parents' lifestyle allowed her to travel at a young age sparking a lifelong curiosity and interest in other cultures. Her grandparents taught her to appreciate nature and our planet earth. Her brother and sister have supported all of her pursuits in multiple ways. A final thank-you goes to Dr. Rajaram for his vision and persistence.

Foreword

Writing is hard even for writers who do it all the time. My entire career has been dedicated to public purpose, from the time I took the Hippocrates oath as a doctor to the time I first initiated a trust for education for paramedics and such others, I have hoped all these years to have the time to pause and reflect on the numbers of things we need to enable, empower our youth to simple action on the ground, and when I see any work that instills such thinking, I stop to admire.

When Dr Rajaram invited me to read this book, I realised that his is an attempt to connect the dots to all that affects us as people in everyday life.

In my work I meet many who work on science and on technology as well as run colleges for the young. Shaping the minds of our young is a process that needs all the help and support we can give. Even more so for the underprivileged who don't have the benefit of good education beyond schooling.

Environment and Climate Change is a step in that direction. As a book, it can help any graduate of basic sciences or humanities or commerce or engineering.

Karnataka has the pride of place today as a pioneer in education reforms and the very first state to have implemented the New Education Policy. The state is setting an example by entirely overhauling the system through various reforms: a single regulator to oversee higher education; no more MPhil courses before PhD; fixed fees for public and private institutions; students can choose between three and four-year undergraduate courses; multiple entry and exit points in degree courses, and many other such initiatives that puts higher education within the reach of many more hundreds of thousands of people.

It is in this larger context that a book like this, that offers lucid examples of how to make ourselves future-ready on the serious concerns of climate change, becomes extremely important. I would urge every faculty member of every college and University and of course every student across disciplines of learning to read and absorb the basic principles that Dr Rajaram has laid out in an easy-to-read-and-practice manner.

India should aim, just as we do, in our humble way in Karnataka, to increase the Gross Enrolment Ratio in higher education including vocational education. Books such as this will be an important tool as we move from 26 percent enrollment in 2018 to 50 percent by 2035.

Even if I pursue a vigorous public career today, my heart lies in healing and in exciting the interest of young minds in creating solutions and opportunities for growth that balances the crisis on the climate front.

Over my years if there's anything that has stayed as a lasting value it is giving; touching others' lives, expanding the circle of our concern to include others. It is this care and compassion that the book imbues. Here's wishing the book every success.

Dr C N Ashwath Narayan
Hon'ble Minister of Information Technology - Biotechnology, Higher Education, Science and Technology of Karnataka [august 2019-to date]; Minister of Skill Development and Entrepreneurship & Livelihood of Karnataka [february 2020-to date]. Former Deputy Chief Minister of Karnataka [august 2019 to july 2021].

Preface

I have felt that the disastrous consequences of Climate Change and the resulting climate extremes should be known by all high school and college students so that they can better manage these extremes and become leaders in fighting Climate Change. I and my co-authors, Ms. Lynn Tiede a teacher in New York City who has been teaching children for over 25 years, and Mr. Keith Olson, a retired teacher who has taught science in high school in Chicago for over 35 years, have written this book to accomplish this purpose. I am an environmental consultant, both in the United States and India, for over 40 years and have seen the disastrous effects of flooding and droughts. This book contains many topics related to Climate Change and many of these are recommended for high school students in India. The unique aspect of this book is the many practical activities and exercises that have been provided at the end of each chapter to challenge the students to learn by doing. This kind of learning is imprinted in their minds and they will become leaders who change society and the government by insisting that we move away from fossil fuels that cause global warming and the resulting climate extremes. Our energy requirements can be met by renewable energy and new sources of energy that the students will innovate when they graduate from college. Many topics on natural resources, biodiversity and ecosystems, energy, natural hazards and disaster management, and managing environmental crises give the student an overview of the impacts of Climate Change on our lives.

There are many technological innovations and solutions available to combat climate change. In addition, society can change their priorities from consumption to conservation to manage climate change. All these are detailed in the book. International and Indian case studies highlight the problems created by humans by focusing on development at any cost attitude instead of a balance between development and environmental protection. As Gandhiji said, the earth is borrowed by the current generation and has to be maintained and improved for the benefit of future generations. The silver lining in the dark clouds of climate extremes are the actions taken by young people around the world to demand action on Climate Change. Greta Thunberg of Sweden and Vinisha Umashankar of India are examples of such young leaders demanding action from local, national and international agencies. They are showing by example what can be done to mitigate the impacts of climate change.

In 2021, I had the good fortune of e-meeting Dr. Hari Haran Chandra of AltTech Foundation of Bengaluru, India. I introduced him to Mary Conley Eggert, Founder of Global Water Works (GWW) in Chicago. GWW and AltTech Foundation have been working on Water Conservation and Recycling and Reuse of Water in our cities through the World of Water Action Forums. We also conduct online Workshops for citizens and experts on various topics related to water. Dr. Hariharan has arranged talks for me at many forums in India, including the Prem Jain Memorial Trust (PJMT) and the Indian Green Building Council (IGBC). These forums have convinced me that citizens are ready for action to combat and mitigate climate change, and this book will help these and many such organizations spread the word further to all citizens in India and around the world. Both PJMT and IGBC are pioneering organizations working with the government and civil society to bring awareness of the solutions available to manage climate change, and I will work with such organizations in their efforts. Working together across international boundaries and educating our youth is the only way to prevent further disastrous impacts on property and loss of life due to the climate extremes that climate change is bringing about. By working with vulnerable populations and making them resilient to climate change is the challenge of the new generation, and this book will help the students in this important work.

I hope that this book will transform the thinking of students and make them aware of the urgent need to change our attitude to development, and work toward a healthier planet for all citizens of this world.

Vasudevan 'Raj' Rajaram
April 2022

Summary

Climate Change and Environment is a book that is written by three professionals in the United States who have extensive experience in environmental issues in India and the United States. It has a lot of hands-on exercises and activities that the teachers and students can use to make the theory of Climate Change and Environment become practical in their day-to-day lives. It covers various topics related to impacts of Climate Change which include Natural Resources, Energy, Ecosystems, Biodiversity. Natural Hazards and Disaster Management and Environmental Crises. It offers solutions to the Impacts of Climate Change such as the Social Response to the Impacts, and Technologies available to combat Climate Change. Several international case studies are presented to illustrate the past environmental crises and how humans have dealt with them. The book lays out a strategy for the government, civil society and students to work toward reducing greenhouse gases, and conserve our natural resources by wise consumption. The role of young people in combating climate change is detailed.

About the Book

This book provides a broad spectrum of facts and figures related to climate change and its devastating impacts involving flood, fire and related disasters. It vividly covers the complex changes occurring in the environment due to global warming caused by greenhouse gas emissions. A striking example that should wake us all up are wildfires during September-2020 in California, Oregon, and Washington States of US. This book underscores the need to reverse the increasing trend of the global warming and elaborates on some of the actions that we must take now to save us from more devastation and destruction that would otherwise frequent the earth in this decade and more so in the future. The urgent need for renewable energy is emphasized. There is useful information on water pollution, forests and minerals in India. This will be a valuable book in the hands of young students who need to understand why we need to act now to reduce these emissions drastically and work towards decreasing the global temperature by resorting to renewable energy. It has case studies on how different projects and movements in India and around the world are making the environment better. These are good examples to follow. Activities and exercises are well covered to reinforce proper learning all throughout the book. I wish this book great success in making climate change a priority for all citizens of the world.

Subijoy Dutta, P.E.,
Founding Director, Rivers of the World Foundation,
Maryland USA.

Chapter 1

Climate Change and Environmental Management

Introduction

The interest in environment is increasing since the scarcity of fresh water, fresh air and land for growing our crops is affecting a large part of the world population. The drastic weather patterns we are experiencing around the world are resulting in a big emphasis on the impact of climate change on humans and other living beings on earth. The authors have written this book to address these topics and provide a practical framework for high school students to understand these complex changes occurring in the environment, and become responsible citizens who can ensure sustainability and resilience in our cities and rural communities.

Since the dawn of the industrial revolution in the 1800s, the accumulation of carbon-dioxide and other greenhouse gases has been increasing at an alarming rate. These gases create a thin film around the earth's surface and trap the heat inside our planet

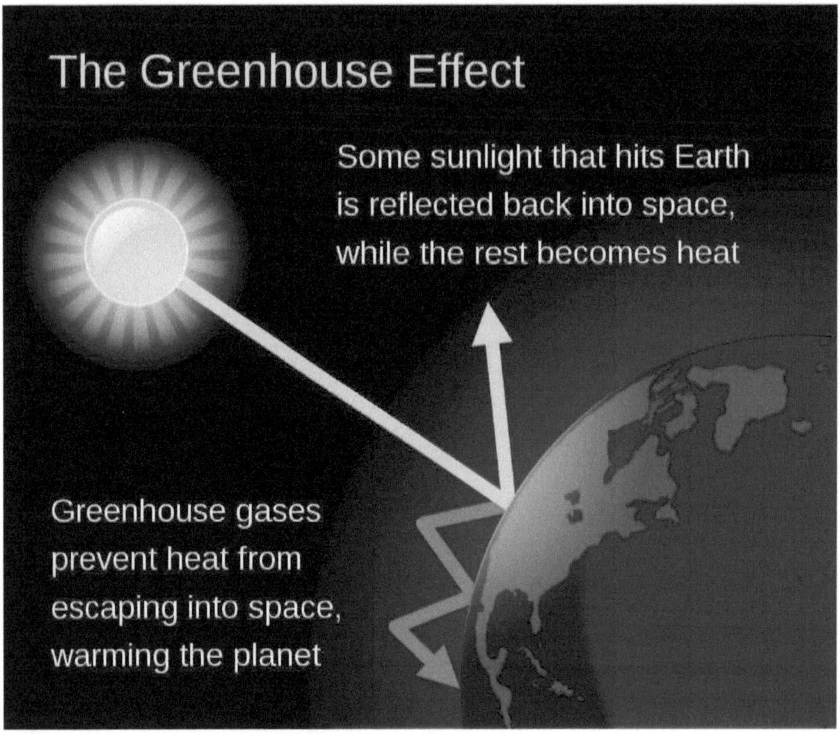

Figure 1.1: ***Greenhouse gases.***

This extra heat trapped in the planet is increasing the temperature in the summers and creating drought conditions around the world. The increase of our temperatures is warming the oceans and this in turn, is creating violent, damaging storms around the world. This caused 200 world leaders to meet in Paris in 2015 and draft an agreement to limit the increase of temperatures around the world to less than 2 degrees Centigrade (c). This book will describe the various ways communities around the world are coping with the climate increases, and taking steps to decrease the accumulation of greenhouse gases in our environment.

The impact on our natural resources due to climate change is felt acutely among the 2 billion people who are living on less than Rs. 150 per day. Droughts are killing many who are living marginalized lives, and access to clean drinking water is unavailable to large segments of the population in Africa, Indian subcontinent, and many other parts of the world. Failure of crops in drought affected areas is forcing many to commit suicide or abandon their farms and move to already crowded cities. Fossil fuel plants are continuing to grow and add more carbon in the atmosphere. The Paris Climate Change agreement is helping to increase solar and wind power but the pace of these projects is still slow. Our forests are shrinking and these further aggravate the accumulation of greenhouse gases.

The rate of change of carbon dioxide (CO_2) concentration in the atmosphere has been accelerating since the dawn of the industrial revolution. Since 1958, measurements of the level of global CO2 have been made and the results of the measurements are depicted in the Keeling Curve, named after Ralph Keeling who started the effort in the US. In May 2018, the level of CO2 passed the 410 parts per million (ppm), for the first time in 800,000 years. Climate change experts have been predicting disastrous effects of climate change at a level of 350 ppm. At the current level of 410 ppm, we are really in trouble with drastic flooding or serious droughts in many parts of the world.

Natural Hazards and Disaster Management

The frequency of natural hazards is increasing worldwide, and the damage due to these disasters in terms of lives lost and property damage is increasing. Experts are of the opinion that the increase of CO2 and the resulting warming of the atmosphere is responsible for the violent storms and forest fires that are happening worldwide. For example, the forest fire season in California, USA, was only in the summer months but now is almost round the year. The excessive heat and drought are main causes for the dry timber in the forests, and this leads to severe fires. The fires in Indian forests are also having a lot of adverse impact and managing these fires takes a lot of resources that are sorely needed for other aspects of our economy.

Environmental Crises due to Climate Change

The impacts of Climate Change are felt most by the vulnerable populations who live in slums or poor rural communities. A hot summer, lack of rainfall or flooding, and other natural crises kill a lot of poor people. The Government of India has taken several steps to avoid such environmental crises and set up several national missions. These are:

1. Solar Mission;
2. Enhanced Energy Efficiency Mission;
3. Sustainable Habitat Mission;

4. Water Mission;
5. Sustaining Himalayan Ecosystem Mission;
6. Green India Mission;
7. Sustainable Agriculture Mission; and
8. Strategic Knowledge Mission for Climate Change.

This is an exciting part of the National Action Plan on Climate Change created by the Indian Ministry of Environment and Forests and Climate Change. We will discuss details of this plan and the various missions in the book.

Impacts on Populations and the Economy

Climate Change is constantly impacting the people and the economy of a country, through the occurrence of hurricanes (also called cyclones or typhoons), floods, fires and droughts throughout the year. Some wealthy economies have spent a lot of money to build infrastructure to mitigate the impacts of Climate Change but these are not sufficient to withstand the constantly increasing ferocity and frequency of the disastrous events. Plans for prevention of these disastrous events have to be put in place and implemented effectively. It should be remembered that the impacts are mostly on the poor communities who are already living marginalized lives and have no ability to cope with these natural disasters which are caused by human actions.

This book is focused on prevention of drastic impacts of Climate Change and what steps can be taken to mitigate their impacts when they do occur. It gives students various activities to analyze the science behind Climate Change and provides a lot of exercises and problems for independent study and understanding. By participating in these activities and exercises, it is hoped that they will become responsible citizens of the future and choose careers in planning for alleviation of Climate Change impacts, and implementing the various programs being undertaken by the Indian Government.

Chapter 2

Climate Change and How to Minimize It

Introduction

Climate change refers to the changes in annual weather patterns caused by global warming. As explained in Chapter 1, global warming occurs as the levels of carbon dioxide in the atmosphere have been increasing dramatically since the inception of the industrial revolution in the 1800s. The temperature in the oceans and land have been increasing omeasured occurred between 2000 and 2017, with 2016 being the hottest. This has resulted in more droughts, fires, floods, mudslides and storms (*Gore, Al*, 2017).

Climate change has received world attention since the Kyoto Protocol of 1997, where Governments from around the world signed a protocol that acknowledged the disastrous impact of climate change on people and economies and decided to take steps to minimize the impact. However, for the world and the United Nations to step up to the challenge and mobilize international support, data was needed to convince the decision makers. Since 1958, measurements have been kept of the level of global carbon dioxide by Ralph Keeling in Hawaii.

A Keeling curve is shown in Figure 2.1.

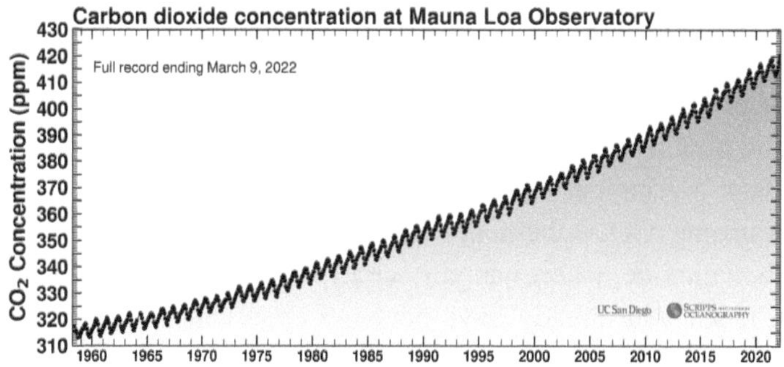

Figure 2.10: ***A Keeling curve depicting the level of global carbon dioxide.***

The trend of the curve is alarming and shows that the level of carbon dioxide in the atmosphere passed 410 parts per million (ppm) in 2018, the first time in 800,000 years (Gore, Al, 2017, pp. 38-39).

In the industrial age since 1800, the amount of carbon dioxide in the atmosphere has increased from about 250 ppm to the current reading over 400 ppm. In May 2018, the level of carbon dioxide in the atmosphere passed 410 ppm for the first time in several hundred thousand years. What does that mean to you and to the world? And who said that, and how do they know? These are questions you can try to answer, starting with some background.

> First, where do the numbers come from? Beginning in 1958, the level of atmospheric carbon dioxide has been measured regularly. Because Charles Keeling started the effort, the graph of the level of carbon dioxide was named after him.
>
> Second, notice the zig-zag shape and the trend of the curve. The yearly zig-zag results from the effect of plants and people. In summer, in the northern hemisphere, plants are growing and taking in CO_2, and the level drops. In the northern winter, plants are dormant and human activity, burning fossil fuels etc., dumps CO_2 into the atmosphere.
>
> The multi-year trend is always up. Starting before the industrial age, atmospheric carbon dioxide was about 250 ppm. When Keeling began his measurements in 1958, it was 315 ppm and since then has passed 400 ppm. The rate of increase is also going up, from less than 1 ppm addition per year in the early 1960's, to recent annual increases between 1.5 and 2.5, even as much as 3 ppm in 2015.

2.1 Basic knowledge of Carbon dioxide

Fossil fuels like coal and petroleum contain carbon, and when they burn, carbon dioxide is produced. As carbon dioxide is the major culprit in global warming, we need to know something about it. This section will cover the properties of carbon dioxide as a molecule, where it comes from, where and how it can be used and some effects of the gas. Of course, the main effect, a principal topic in this book, is global climate change and global warming. CO_2 is the major cause of climate change.

Carbon dioxide, where can you see it? As it is an odorless gas, usually you can't, but you already have personal experience. Certainly, you've drunk a carbonated drink, whether Coca Cola or a local brand. The bubbles? It is carbon dioxide.

Here's a demonstration of how you can produce those bubbles safely. You'll need only a few supplies:

- Baking soda (sodium bicarbonate)
- Vinegar
- A spoon and a bowl

These are harmless chemicals and are used in recipes.

Procedure:

Place the bowl on a plain surface, put a spoonful of baking soda into the dry bowl. Add two spoons of vinegar and notice what happens.

You'll see bubbles, it is carbon dioxide.

In words, the equation for the reaction is:

Sodium bicarbonate plus acetic acid yields sodium acetate plus carbonic acid.

Using chemical formulas, the equation is:

$$NaHCO_3 + HAc \rightarrow NaAc + H_2CO_3$$
$$H2CO_3 \rightarrow H_2O + CO_2$$

Some things to note:

In the first equation, acetate in the acetic acid is simplified to Ac. Don't confuse with Ac as a symbol of the element actinium. The CO_2 in the second equation is the source of bubbles in carbonated drinks.

Carbon dioxide is part of your life as with each breath you take, you exhale CO_2. The C in CO_2 comes from carbon in the food you eat, whether protein, fat and oil, or carbohydrates.

Demonstration,

Supplies:

- Lime water (CaO in water)
- Carbon dioxide (you'll supply that)
- a straw

If your school has some lime water, blow through the straw into the lime water. The lime water turns cloudy from combining with the CO2 in your breath producing calcium carbonate, $CaCO_3$, which isn't soluble.

The reaction is: $CaO + CO_2 \rightarrow CaCO_3$

Carbon dioxide (CO_2) is heavier than air. You can show that with this simple demonstration.

Supplies:

- Baking soda (sodium bicarbonate)
- Vinegar
- a spoon and a jar

Mix baking soda and vinegar as you did before, only this time in a jar. Use several spoons of baking soda and add twice as many spoons of vinegar. Have the jar on a towel in case it bubbles over. After the bubbling stops, tilt the jar over a candle flame, carefully so that the solid and liquid remain in the jar. Even though you can't see it, you are pouring a stream of CO_2 out of the jar. The candle will go out as the CO_2 cuts off oxygen from the flame.

Although you've demonstrated that CO_2 is heavier than air, how much heavier is it? The weight of air is an average between the weights of nitrogen (28) and oxygen (32), or about 29. Note that the weights of nitrogen and oxygen are double what you see on a list of the elements, 14 for nitrogen and 16 for oxygen. Both elements are diatomic, that is two atoms per molecule, N_2 and O_2. If there were equal amount of both gases in air, the average would be 30, but there's more nitrogen, lowering the average. Those two gases make up about 99% of the dry air, the rest is mostly Argon, plus some CO_2 and trace amounts of other gases, but those small amounts don't change the average very much. Now CO_2. Add 12 for the carbon and 32 for the two oxygens for a total of 44, clearly heavier than the 29 for air.

When the candle went out, you witnessed that CO_2 is heavier than air, but you may wonder why it doesn't settle out the atmosphere, forming a layer on the ground. That gases mix easily gives us the explanation. You are aware of the mixing with cooking odors, you can smell those at distance because the food odors spread through the gases in the atmosphere. Likewise, CO_2 mixes with nitrogen and oxygen and the percentage of each gas doesn't vary with altitude.

The density of CO_2 is useful for putting out fires. The CO_2 from a fire extinguisher will settle somewhat, smothering a fire just as you demonstrated with CO_2 poured on a candle.

If you and a responsible adult can work a CO_2 fire extinguisher, you can see the smothering in action. Perhaps, in a safe way and a safe place, you can put out a small test fire. Even without putting out a fire, you'll enjoy as well as learn from practice with the extinguisher.

Your experience will include things you probably wouldn't expect:

- it's noisy
- there's a cloud
- there's a lot of pressure
- there may be bits of white solid

When CO_2 comes out of the extinguisher, it expands rapidly, a cooling process. The cooling is enough to turn some of the CO_2 solid, but the cloud is mostly water vapour in the air, chilling and condensing. While you're now aware you can't see carbon dioxide as a gas, how about as a liquid? That's the way it is in the fire extinguisher. At room temperature, high pressure is essential to keep CO_2 liquid -the sturdy cylinder of the extinguisher withstands the pressure. Squeeze the handle, the pressure is released, rapidly forcing out the CO_2.

You can see solid carbon dioxide, commonly called dry ice, it's called that because it doesn't turn to liquid. Normal ice, H20, goes from ice to water to steam, that is from solid to liquid to gas,. At room pressures, carbon dioxide skips melting and evaporation, going directly from solid (dry ice) to gas, a process called sublimation.

Have you ever been in a cave? A well known cave in India is Ajanta but that is carved into rock. For our subject of climate change, a better choice would be Gupteswar in Orissa. What possible connection could there be between caves and climate change? The link is CO_2.

Gupteswar and other caves formed in limestone when the slow dripping of water containing carbonic acid gradually ate away at the stone. The dissolved limestone washes away leaving caverns, some massive.

In words, the equation is: limestone (calcium carbonate) plus carbonic acid yields calcium hydrogen carbonate.

Rewriting that with chemical symbols gives:

$$CaCO_3 + H_2CO_3 \rightarrow Ca(HCO_3)_2$$

While calcium carbonate ($CaCO_3$) doesn't dissolve in water, calcium hydrogen carbonate ($Ca(HCO_3)_2$) does, and is washed away. Calcium hydrogen carbonate can also be called calcium bicarbonate. The same reaction is the cause of deterioration of some statues and building stones which, like caves, are limestone and marble, a harder form of $CaCO_3$. Perhaps you know or can learn of other caves found in limestone in India and other countries. Regardless of the geology, many caves are tourist sites, well worth visiting.

As carbonic acid, the combination or carbon dioxide, is the active chemical for cave formation, it is also critical for the deterioration of coral reefs. Much as carbonic acid degrades statues, it degrades reef coral. Equally, or maybe more importantly, as the climate warms, oceans warm. As corals thrive in only a narrow temperature range, when the temperature changes, they are doomed. As the corals die off, the remains look white, hence the term Coral bleaching, which is happening worldwide. Two thirds of the largest and best-known reef system in the world, the Great Barrier Reef off the east coast of Australia, was dead or severely bleached by 2017. Many corals around India fare no better. For instance, death is likely for the corals of the Maldives. As an activity, see what you can learn of the health of corals worldwide, but particularly near India.

Let's return to the source of the carbon dioxide, fossil fuels. Fossil fuels get that name from being buried for millions of years, the remnants of plants 'fossilized' in those eons. All of them, coal, oil, petroleum are chemically primarily carbon and hydrogen. Coal is mainly carbon and when it burns, that is combined with the oxygen in the air, carbon dioxide is produced. The equation is: $C + O_2 \rightarrow CO_2$

Other than elemental carbon in coal, the simplest fossil fuel chemically is methane, CH_4, often called natural gas, which has one carbon atom per molecule. Writing the equation of burning methane illustrates the products of combustion.

Methane plus oxygen produces carbon dioxide plus water

$$CH_4 + 2\,O_2 \rightarrow CO_2 + 2\,H_2O$$

As noted above, CO_2 is colorless and odorless, thus hard to detect. The water shows up in a way you can witness. If you put a pan on a stove above the flame, you'll notice a film on the pan. It doesn't last long, it evaporates when the flame heats the pan, but that film is water from the burning of the fossil fuel. The water vapor becomes visible as the film on the pan.

You probably have used other fossil fuels or are familiar with them. Propane, with three carbon atoms, is, like methane, a gas, stored and sold in sturdy containers. Butane has four carbon atoms; octane has eight and so on. Coal and the others when burned, whether for power or transportation or heat, all produce carbon dioxide and water. Burning them also produces pollutants, like ash and sulfur dioxide, in addition to the carbon dioxide responsible for climate change.

2.2 Paris Agreement

The Paris Agreement was reached by the leaders of the United States (US), India, China, Canada and the European Union (EU) in 2015 in Paris, France. It is a major landmark in the effort to contain global carbon dioxide emissions in the countries of the world. Many countries have adopted green technologies such as solar, wind and bioenergy since the signing of this agreement. It is an ongoing effort with countries reporting their carbon dioxide emissions at predetermined dates and efforts taken to reduce them.

Global carbon dioxide emissions by jurisdiction for 2018 are (Wikipedia, July 26, 2018):

Table 2.1: *Global carbon dioxide emissions in 2018*

Country	CO_2 emission (in percentage)
China	29.4%
USA	24.3%
European Union	9.5%
India	6.8%
Russia	4.9%
Japan	3.5%
All other countries	21.6%

Since Kyoto, there have been many attempts to improve the global agreement and the Paris agreement of 2015 was a tremendous achievement by over 200 countries. The highlights of the Paris Agreement is summarized below (Paris Agreement, Wikipedia, July 26, 2018).

- 196 countries (called State Parties) negotiated at the 21st Conference of Parties (COP21) of the United Nations Framework Convention on Climate Change (UNFCCC) in Le Bourget, near Paris, France;
- The agreement was adopted on 12 December 2015, and is subject to review every 5 years, starting in 2023;
- As of July 2018, 194 states and the European Union (EU) have acceded to the agreement, and represent more than 87% of the global greenhouse gas emissions. These include China, India and the United States, the countries that emit about 60.5% of the greenhouse gas emissions. However, the US withdrew from the agreement but many communities and companies in the US are living by the terms of the agreement;

- The agreement holds the increase in global average temperature to well below 2 °C above pre-industrial levels and wishes to pursue efforts to limit the temperature increase to 1.5 °C above pre-industrial levels;
- It increases the ability to adapt to the adverse impacts of climate change and foster climate resilience and low greenhouse gas emissions development, in a manner that does not threaten food production;
- It makes finance flows consistent with a pathway towards low greenhouse gas emissions and climate resilient development;
- It aims to reach "global peaking of greenhouse gas emissions as soon as possible". It is an incentive for and driver of fossil fuel divestment.

A key provision of the Agreement is called "Nationally Determined Contributions (NDCs)". This is a contribution that each individual country should make in order to achieve the worldwide goal that is determined by all countries individually. Article 3 requires them to be ambitious, represent a progression over time, and set with the view to achieving the purpose of this agreement. The contributions should be reported every 5 years and are to be registered with the UNFCCC Secretariat. Each further ambition should be more ambitious than the previous one, known as the principle of Progression (Wikipedia, July 26, 2018). Countries can cooperate and pool their NDCs. There is no enforcement mechanism for NDCs, but as Janos Pasztor, the UN Secretary General on Climate Change mentioned in Paris, only a "name and encourage plan".

The NDC Partnership (https://ndcpartnerships.org/partners) was launched at COP22 in Marrakesh to enhance cooperation so that countries have access to the technical knowledge and financial support they need to achieve large-scale climate change and sustainable development targets. The NDC Partnership is co-chaired by the governments of Germany and Morocco and includes 71 member countries, 13 institutional partners and two associate members.

2.3 Effects on Global Temperature

The negotiators of the agreement stated that the NDCs and the 2 °C reduction were insufficient, instead, a 1.5°C is required. This requires the projected level of aggregate greenhouse gas emissions to be 40 gigatonnes instead of the expected 55 gigatonnes in 2030. In the first half of 2016, average temperatures were about 1.3 °C above the average in 1880, when global record keeping began. In October 2016, US President Barack Obama claimed that "Even if we meet every target .... we will only get to part of where we need to go". He also said that "this agreement will help delay or avoid some of the worst consequences of climate change. It will help other nations ratchet down their emissions over time, and set bolder targets as technology advances, all under a strong system of transparency that allows each nation to evaluate progress of all other nations".

The implementation of the agreement by all member countries together will be evaluated every 5 years, with the first evaluation in 2023. The outcome is to be used as input for new NDC of member states. This stocktake will not be of contributions/achievements of individual countries but a collective analysis of what has been achieved and what more needs to be done. It will also evaluate adaptation, climate finance provisions, and technology development and transfer. The objective of the stocktake is to evaluate how the new NDCs must evolve so that they are ratcheted up and reflect a country's "highest possible ambition".

Structure of the Agreement

A key difference between the Paris Agreement and the Kyoto Protocol is their scopes (Wikipedia, July 26, 2018). Under Kyoto, two separate groups of countries (developed and developing) had different responsibilities whereas Paris requires all countries to submit emission reduction plans. The Paris agreement has a "bottom up" structure unlike most international environmental law treaties. While Kyoto sets commitment targets which have legal force, Paris emphasizes consensus building and allows for voluntary and nationally determined targets. The agreement is considered an "executive agreement rather than a treaty", and can be considered a framework for a "global carbon market".

Mitigation Provisions and Carbon Markets

Article 6 of the Agreement establishes a framework to govern the Internationally Transferred Mitigation Outcomes (ITMOs). Provisions of the Article require the "linkage of various carbon emissions trading systems to avoid double counting, transferred mitigation outcomes must be recorded as a gain of emission units for one party and a reduction of emission units for the other. The ITMO will provide a format for global linkage under the auspices of the UNFCCC. This provision creates a pressure for countries to adopt emission trading systems and to monitor carbon units for their economies.

Paragraphs 6.4 to 6.7 establish a mechanism to "contribute to the mitigation of greenhouse gases and support sustainable development". The Sustainable Development Mechanism (SDM) is considered to be the successor to the Clean Development Mechanism established under the Kyoto Protocol. The goals of the SDM are:

1. To contribute to global GHG emissions reductions; and
2. To support sustainable development.

The SDM will be available to all parties so that cooperation among countries can happen. The key is to set the price of "Certified Emission Reductions" in such a way that it will create demand for projects focused on GHD reductions and sustainable development in all countries.

Countries must also report on their adaptation actions, making adaptation a parallel component of the agreement with mitigation. The adaptation goals focus on enhancing adaptive capacity, increasing resilience, and limiting vulnerability.

Financing for Making the Agreement Work

Money is needed for implementing Green Technologies and also technology transfer amongst countries helps new green technologies become economically feasible sooner. As the Paris agreement was being negotiated, developed countries reaffirmed their commitment to mobilize $100 billion a year in climate finance by 2020, and agreed to continue mobilizing finance at the $100 billion level till 2025.

Though both mitigation and adaptation require increased financing, adaptation has typically received lower levels of support and has mobilized less action from the private sector (Wikipedia, July 26, 2018). The Paris agreement called for a balance of climate finance between adaptation and mitigation, and specifically underscored the need to increase adaptation support for parties most vulnerable to the effects of climate change, including the Least Developed countries and Small Island Developing States. The agreement reminded governments of the importance of public grants, because adaptation measures receive less investment from both public and private sectors. John Kerry of the United States announced that the US would double its grant based adaptation finance by 2020.

Some specific outcomes from the Paris agreement are:

1. US $420 million for Climate Risk Insurance;
2. Launching for a Climate Risk and Early Warning Systems (CREWS) Initiative;
3. Launching of a Green Climate Fund, which has received pledges for over $10 billion. Pledges are from both developed (France, US and Japan) and developing countries (Mexico, Indonesia and Vietnam).

2.4 Loss and Damage due to Climate Change

The Alliance of Small Island States and the Least Developed Countries, whose economies and livelihoods are most vulnerable to the negative impacts of climate change, pushed to address the Loss and Damage resulting from climate change. This is because many of the worst effects of climate change will be too severe and come too quickly to be avoided by adaptation measures. In the end, all parties acknowledged the need for "averting, minimizing, and addressing loss and damage" but any mention of compensation liability is excluded. The agreement adopted the Warsaw International Mechanism for Loss and Damage, an institution that will attempt to address questions about how to classify, address and share responsibility for loss and damage.

A major achievement of the Paris agreement is that the Parties are legally bound to have their progress tracked by technical expert review to assess achievement toward the NDC, and to determine ways to strengthen ambition. Article 13 of the agreement articulates an "enhanced transparency framework for action and support" that establishes harmonized monitoring, reporting and verification (MRV) requirements. Thus, both developed and developing nations must report every two years on their mitigation efforts, and all Parties will be subject to both technical and peer review. The framework provides "built-in flexibility" to distinguish between developed and developing countries' capacities. The agreement develops a Capacity Building Initiative for Transparency to assist developing countries in the building of necessary institutions and processes for complying with the transparency framework.

Implementation

The process of translating the Paris agreement into national agendas and implementation has started. The Least Developed Countries (LDCs) have developed Renewable Energy and Energy Efficiency Initiative for

Sustainable Development, known as LDC REEEI. This is set to bring sustainable and clean energy to millions of energy starved people in LDCs, facilitating improved energy access, the creation of jobs and contributing to the achievement of the Sustainable Development Goals.

India's efforts to Implement the Paris Agreement

India has been aggressive in managing Climate Change and changed the name of the Ministry of Environment & Forests to the Ministry of Environment, Forests & Climate Change in 2014. In 2009, the Indian Government launched the National Action Plan for Climate Change (www.moef.nic.in/downloads/home/pg.01-52.pdf). As mentioned in Chapter 1, eight missions were launched under this plan and all of them will lead to minimization of greenhouse gases and adaptation to climate change. Although India's contribution to the historical cumulative emissions is around 3%, India's submissions to the UN's NDC focuses on clean energy, with a huge emphasis on solar and wind, and the planting of more forests by 2030 to absorb carbon emissions. India's Prime Minister Mr. Narendra Modi has chartered a new direction on both energy policy and climate change, establishing new political priorities and re-directing investment. Although domestic economic development remains a core priority, the Modi administration is focusing on up-scaling clean energy production to help drive growth while also shifting the country's stance in international negotiations on climate change (www.climatenexus.org/climatenewsarchive/Indiaandclimatepolicy/).

Salient features of India's Climate Policy are summarized below.

- Doubling the coal tax to fund clean energy, from Rs. 50 to Rs. 100/tonne.
- Solar for every home by 2019. The Administration is promoting very large solar power plants, and has invested in 25 solar parks, which could increase the total installed solar capacity ten-fold. It is actively recruiting multi-billion-dollar funds in addition to allocating $81 million in the current budget.
- The government is preparing to double wind power to 60,000 MW by 2022, and has set a goal of 100 GW for solar power. SunEdison has announced US $4 billion for a large solar panel factory in India.
- Investing in Clean Power, with $100 billion in renewables, $50 billion in transmission and distribution and $100 billion in all clean coal and

nuclear power. 80% accelerated depreciation allowance will be set for renewable energy.
- Subsidizing efficient LEDs in 100 cities, with a subsidized price of Rs. 130 compared to the market price of Rs. 600 per LED.
- Stopping imported thermal coal, with renewables and energy efficiency as priorities.
- Setting up a National Adaptation Fund of $18.5 million, with a focus on agriculture.
- Increased global climate cooperation, with full support for the Paris Agreement.

In addition to the actions by the Central government, state governments are taking their own actions to mitigate and adapt to climate change. New Delhi and neighbouring Haryana states are mandating solar rooftop installations for buildings over 500 square yards. States are taking steps to promote waste to energy, and solar and wind programs in their states. The State of Gujarat is promoting wind energy and working with industry to increase renewable energy projects.

The increasing use of renewable energy around the world gives a glimpse of hope that we can limit the temperature increase by 2050 to under 2 °C. Without the reduction of greenhouse gases (carbon dioxide, methane being the big ones), the world will face the following consequences:

1. Violent storms that result in floods that damage property and lives, especially in the coastal areas around the world;
2. Extreme heat and resulting droughts in many parts of the world which will reduce food production and result in the loss of thousands of lives in the cities and rural areas around the world;
3. Dangerous fires as were witnessed in California, Australia and many parts of the world between 2015 and 2020. These fires result in loss of property, human lives, and wildlife in our forests, and when these forests are burned down, temperatures will further increase due lack of vegetative cover;
4. Melting of the glaciers and ice sheets in the Arctic and Antarctic Oceans, resulting in flooding and sea-level increases in many coastal regions around the world. This has the potential to result in trillions of dollars' worth of damage to property and loss of thousands of human lives; and

5. Damaging mudslides due to the loss of vegetation in mountainous areas. This will result in damage to property and human lives in the mountainous regions of the world.

The implementation of the Paris Agreement and steps taken by global corporations and governments should limit the temperature increase to less than 2 °C and minimize the adverse impacts mentioned above. Students of the coming generation will have to take an active role in advocating for policies that reduce the temperature increase, and this book should provide a lot of ideas for taking on this active role and saving the planet from imminent danger.

2.5 Activities and Exercises, Field Observation and Project

1. Explain the Keeling Curve to someone, a classmate, your parents,
2. Explain the greenhouse effect to others, including why carbon dioxide is a greenhouse gas.
3. Conduct a poll about CO_2 levels and the greenhouse effect. Design your question, perhaps asking people to say which issue is more important, global climate change or another issue of your choice. Ask the question in your school or neighborhood.
    - When did the carbon dioxide level break 400?
    - Does it show any sign of going back to that level?
    - What level is agreed on in the Paris Accords?
    - How much has the level changed in your lifetime?
    - How much has the level changed in the lifetime of your parents?
4. Extend (extrapolate) the Keeling curve to 2070. What will be the CO_2 level then, assuming no change in the continuing use of fossil fuels? On average people born in India in 2000 will live to the year 2070.
5. 400 ppm sounds like a small number, even more so as a percent, that's only 0.04% of the atmosphere. Why is such a small number so important to climate? (hint, remember it's a greenhouse gas).
6. Imagine you are a reporter for a newspaper or television station. Your editor has just seen a report from the IPCC (International Panel on Climate Change) and assigns you to have a summary ready for the paper or station. Write the text you'd submit. Follow the rules of journalism, including what, who, where, when and why. Know your sources and know whether you are writing just a report or including an opinion.

7. Write a letter to a newspaper about climate change, whether explaining the curve, noting what's happening, or arguing for action.
   The Keeling Curve is based on measurements maintained by the Scripps Institute in California, U.S.A. Their website is: www.scripps.ucsd.edu/programs/keelingcurve
   Additional websites are:
   http://www.esrl.noaa.gov/gmd/ccgg/trends/
   www.en.wikipedia.org/wiki/Keeling_Curve
   www.Kcurveprize.org
   www.350.org
8. The text says science has data for carbon dioxide for 800,000 years. How do scientists know this? While the Keeling curve starts in 1958, researchers have extracted carbon dioxide from ice cores in Greenland and Antarctica, ice that is tens of thousands of years old. You can look in encyclopedias and other sources for the methods of learning about ancient carbon dioxide and illustrate the information to others. Look for the 'hockey stick' model which combines carbon dioxide levels from millennia ago with that from less than one hundred years.
9. Read through the Paris Climate Agreement and watch coverage of the signing. Write a summary of the key provisions and analyze why it was so historical. https://unfccc.int/process-and-meetings/the-paris-agreement/.
10. Analyze how the Paris Climate Agreement as an international agreement has to filter down through civil society to actually work. Look at the mechanisms and agreements that each country must adhere to and find evidence by researching to find out how local governments, organizations, businesses, and individuals are implementing projects in your area to address climate change. Create a graphic organizer in the form of a flow chart to highlight how the global agreement must be implemented on the local scale.
11. Make a timeline of key historical events that led to the increase of greenhouse gases. Find books, articles, and videos that provide background information, take notes and synthesize the information to create a visual of how and why the problem has developed since the Industrial Revolution and spread of Industrialization globally.
12. Assess your own carbon footprint (the amount of carbon you generate). Use a calculator such as this: http://www.meetthegreens.org/features/carbon-calculator.html. Decide upon 5 steps you can take to reduce

your carbon footprint and practice them over a period of a month. Journal and log your efforts. Continue to adopt new practices each month. Continue for as long as you can. Research new ideas continually and keep a file of best practices.

13. Analyze the lifestyles of people in the United States and India to assess their carbon footprints as well as the challenges they are facing due to climate change. Then look up solutions being worked upon in each country. Create a booklet to describe key aspects of carbon footprint, challenges faced, and attempted solutions for people in the United States and India.
14. Imagine a city with a small carbon footprint, and describe the city with respect to the number of trees in various neighborhoods, lifestyle changes made by people to reduce water and energy consumption, and minimizing the generation of plastic and other wastes. How should such a city build relationships with people in various neighborhoods?
15. Study the National Action Plan for Climate Change, and summarize how your city plans to help implement the Plan. What portion of the plan can be implemented in your home if you urge your parents to take the necessary actions?
16. Obtain the latest report from the Ministry of Environment, Forests, and Climate Change and write a 5-page report on actions taken so far by the Ministry and State Governments to implement the National Action Plan for Climate Change.
17. Why do you think the rich nations have to help developing nations and smaller countries to reduce greenhouse gases? What steps, other than providing funds, should the rich nations take to help developing nations and smaller countries to implement the steps necessary to mitigate climate change impacts?
18. Study the increase in renewable energy production (wind, solar and biofuels) by Indian companies since 2015, and write a short report (2 to 3 pages) on what kinds of renewable energy projects have been undertaken by Indian companies (private and public). How has your community changed its energy and municipal services delivery since 2015?
19. How would you advocate for government policies that reduce greenhouse gases? Study the impact that Ms. Greta Thunberg of Sweden has had on global governments and corporations to reduce their greenhouse gases. How would you urge your parents to vote on

climate change issues so that you have a sustainable planet to live in? Write a 5-page report about your advocacy plans.

20. At 400 ppm, the amount of carbon dioxide in the atmosphere seems small and unimportant. Here is a comparison to another number: the permitted limit of blood alcohol for driving is usually given in percent, usually from 0.05% to 0.10%. Change 0,05% to ppm and compare your answer to 400 ppm. Does the 400 ppm still seem unimportant?
21. The atmosphere seems vast but is actually a small layer. On the world's tallest mountain, the air is too thin to sustain human life. That is at an altitude of 9000 meters above sea level. Going horizontally instead of vertically, name a few locations 9 km from where you live and what is the amount of carbon dioxide emission in those places?
22. Very large numbers are hard to comprehend, including gigatonnes of carbon dioxide, the weight discharged into the atmosphere yearly. What does giga mean? Although carbon dioxide is a gas, think of a solid fuel, coal with that weight. How many rail cars would be needed to carry that weight? How long would that train be?
23. The prefix tera is 1000 times as much as giga. Global energy demand is currently about 18 terawatts (TW). Find the energy use of Delhi. How many cities the size of Delhi would use 18 TW?
24. How much carbon dioxide is produced from vehicles? To start, guess the answer to this question: what is the weight of the carbon dioxide exhaust from an auto driven an average distance in a year compared to the weight of the vehicle? The auto could be a Maruti 800 that weighs 650 kg or a vehicle of your choice. An exact answer isn't needed, just an estimate. Perhaps you'd say one tenth of the weight of the vehicle? A way to get the answer is provided below.

Assume the vehicle goes 20 km on a litre of fuel, claimed for the Maruti for highway miles, and it's driven 6000 km in a year. That would use 300 litres, which weighs about 210 kg. Using that much fuel would produce 3 times that weight of carbon dioxide (CO_2), or 630 kg. How does that compare to the weight of the vehicle? For the Maruti 800 it's about the same. Most people's guess is much lower than the actual output – how did your guess compare? The meaning is clear – any driving we do uses the atmosphere as a dumping place for our waste carbon dioxide.

Another question, are these reasonable numbers? What vehicles would do better and worse than 20 km/L? 6000 km a year is 16.4 km per day. Is that a reasonable average distance or too high or too little?

Supporting information and comments:

This calculation gives values that relate to common experience, hence is easier to remember than published statistics.

Assume motor fuel is the hydrocarbon C_8H_{18}, octane, which has a specific gravity of 0.7. It burns according to the equation:

$$C_8H_{18} + 12.5\ O_2 \rightarrow 8\ CO_2 + 9\ H_2O$$

One unit of C_8H_{18} has a mass of 114 grams; one of CO_2 is 44 grams.

As 8 units of CO_2 are produced for every one of C_8H_{18}, 352 grams of CO_2 are produced from 114 grams of C_8H_{18}, giving a weight ratio of 352/114 or 3.09. Rounding the ratio to 3, one kilogramme of petrol when burned produces three kilogrammes of CO_2. By volume, as one litre of fuel weighs 0.7 kilogrammes, the weight of CO_2 produced by burning that volume would be 2.1 kilogrammes.

The ratio doesn't change if kilograms are used instead of grams and is approximate as motor fuel is a mixture of many different types of molecules. Some of those fuels would give higher ratio than three; others a lower ratio, particularly those such as ethanol that already contain oxygen.

One of the isomers of octane (C_8H_{18}), 2,2,4 – trimethyl pentane, is the standard for the octane rating system. A pure sample of this isomer would have a rating of 100 octane.

As most of the petrol burns cleanly, almost all of the exhaust by weight is carbon dioxide and water. Although other products such as carbon monoxide are of major concern as pollutants, their weight percent is small.

25. Trees can take up (sequester) carbon as they grow by removing carbon dioxide from the air and combining it with water to produce cellulose. Working with an 'average' tree, we can estimate how valuable trees are for reducing carbon, whether in town or forest.

To estimate the amount of carbon sequestered by a growing tree, assumptions must be made about the size and growth rate of the tree. For this example, use a tree of 5 metres height and 0.20 metres diameter, which increases its diameter by 2 centimetres in a year with no increase in height. The volume added will be 0.033 cubic metres. Using 500 kg per cubic metre of wood, the tree would have added 16.5 kilogrammes. More details about the calculation are at the end of this exercise.

As a tree is about 50% carbon dry weight, to add 16.5 Kg of wood, the tree would store 8.2 Kg of carbon by using 30 kg of CO_2.

Carbon dioxide has a weight of 44 and carbon 12, a ratio of 11 to 3. Multiply the 8.2 kg by 11/3 gives the 30 kg.

As 1 litre of petrol when burned produces about 2.2 Kg of carbon dioxide, the tree taking up the 30 Kg of would offset the carbon dioxide produced by driving enough to burn about 14 litres.

Does the carbon dioxide exhaust overwhelm what trees can sequester? In a city, even if the trees don't take up all the carbon dioxide, are they still valuable? Would other plants also do well? If crops like maize are plowed under after harvest, the carbon is sequestered for a time. Fast growing plants like bamboo and crop waste could be converted to biochar.

Although the result is small for each tree, one tree in one year taking up the carbon from only 14 L of fuel, it's a positive contribution.

References

1. Gore, Al, 2017, Truth to Power: An Inconvenient Sequel, Rodale Press, 733 Third Avenue, New York, NY 10017
2. https://unfccc.int/process-and-meeting/the-paris-agreement/
3. Ministry of Environment, Forests, and Climate Change, 2012, National Action Plan for Climate Change, www.moef.nic.in/downloads/home/pg. 01-52.pdf.
4. www.climatenexus.org/climatenewsarchive/indiaandclimatepolicy/
5. https://ndcpartnerships.org/partners
6. http://meetthegreens.org/features/carbon-calculator.html.
7. https://en.wikipedia.org/wiki/July 26, 2018.

Chapter 3

Natural Resources

Introduction

The depletion of natural resources due to excessive consumption and waste will have a significant & adverse impact on the planet. As discussed in Chapter 2, we have reached a critical stage with increasing amounts of greenhouse gases in the atmosphere. This chapter will discuss how conservation of natural resources is important to the maintenance of the quality of life for generations to come. Natural resources are the key element in providing the quality of life we enjoy in the 21st century. Water, forests, mineral resources, air quality, and agriculture are the backbone for a thriving community, and it is the government's role to ensure that these are well managed and available for one and all. However, each global citizen has to do his or her part to use them prudently and conserve them so that they are available to present and future generations. Sustainable and effective use of these resources will be described in this chapter. This chapter provides an overview of the water, mineral, forest and land resources available worldwide and in India. The laws affecting the use of these resources in India are briefly described.

Humans are destroying forests to acquire more land for agriculture and urban development. Rapid urbanization is depleting open land and forests and adversely impacting climate change. Mining mineral resources and harvesting forests without adequate environmental protection has long-term effects on the quality of air, land and water. Effective management of these resources is also discussed in this Chapter.

Activities and exercises are provided at the end of the chapter so that the student can do his/her own research and field observations to understand the abundant natural resources available to us and how we can manage them well as global citizens. This will help teachers and students plan independent reading and field projects to appreciate and understand the management of natural resources.

3.1 Water

Water covers about 70% of the earth's surface, but all of this is not suitable for human and animal needs. Only about 3% of the total water is available for drinking and other daily needs. Water is extensively used in agriculture, mining and industrial production, and for human needs. It is also vital to maintaining ecosystems that are essential for our well-being. We have to understand the sources of water, and how to harness them efficiently for various uses. Essentially, water is life.

3.1.1 Hydrologic Cycle

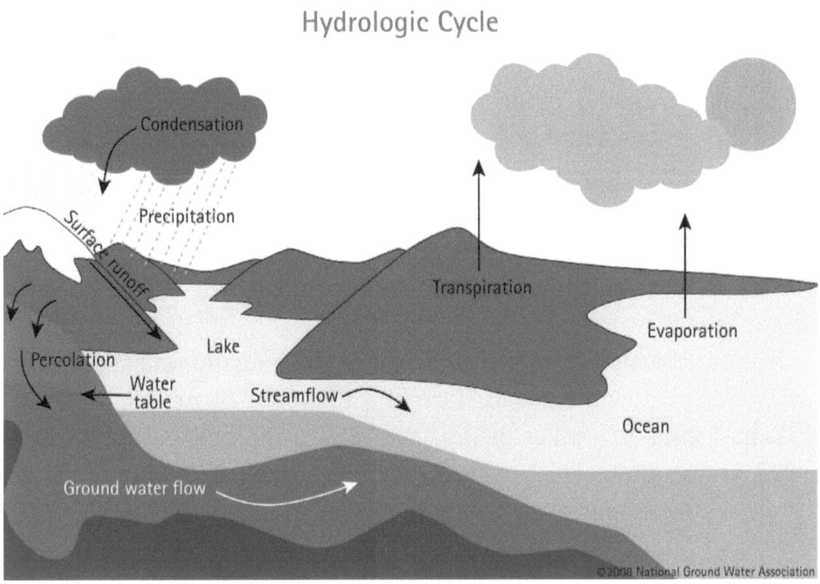

Figure 3.1: ***The Hydrologic cycle.***
Source: https://wellowner.org/resources/groundwater/the-hydrologic-cycle/

It starts with rain which provides the freshwater for our lakes, rivers, and underground aquifers. The hydrologic cycle starts with evaporation of the water in our surface water bodies, including the oceans, and this is stored in the atmosphere. It comes back as rain and the rivers then merge into our oceans and the cycle starts all over again. With climate change on a large scale worldwide, this natural cycle is getting distorted by the oceans warming more than usual and the rain that comes either creates floods in certain places and drought in certain other places. To counteract this adverse impact, we can sequester the carbon in land and in underground caverns (exhaust from fossil fired thermal power plants). Another important and relatively easy method is to recycle the nutrients in treated wastewater to land and enhance the productivity of the land.

3.1.2 Integrated Water Management

Water is the source for all life, from human to animals and plants. We have to learn to manage this well to prevent adverse impacts on humans, animals and plants worldwide. A concept that has been promoted by the United Nations is Integrated Water Management in which water and wastewater are considered resources and are managed efficiently. The water that God has provided us can be used for consumption by plants and animals, and the treated wastewater can be used for agricultural production since this will enhance agricultural productivity. This is due to the nutrients present in treated wastewater. At present, we treat water as a commodity for consumption by humans, animals and plants, and discharge the wastewater into our fresh water bodies in many places where there is no wastewater treatment plant. Even when it is treated in major urban centers, the treated wastewater is discharged to rivers and lakes and then pumped to a water filtration plant for treatment and supply to urban populations.

Integrated water management involves planning the water supply and wastewater treatment infrastructure in a manner that minimizes overall costs, maintains the water quality and allows for efficient recycling of treated wastewater. This requires that the water filtration plant and wastewater treatment plant be located next to each other, with treated wastewater filtered and chlorinated before being pumped to the residents of the city. This will minimize costs and ensure that all the water is optimally used for domestic, industrial and agricultural uses. It will also prevent contamination of water bodies, especially our rivers.

User fees are required to recover the cost of the treatment plants and water distribution system. The user fees are based on the quantity of water used and wastewater generated so that water conservation can be promoted. The weaker sections of society have to be provided water and wastewater treatment at no cost and subsidized by other taxpayers. Many rural communities depend on surface water resources and it is the duty of every citizen and the government to ensure that surface water bodies are not contaminated. In India, many of the rivers are contaminated by industrial and municipal waste and this has a drastic impact on the life of the poor people who depend on this water source. Water borne diseases reduce the life and productivity of the poor people who depend on contaminated water sources for their daily living.

It is a challenge for many municipalities and local governments to build and maintain the water infrastructure. Without the Central and State Government allocating financial and technical resources for water management, the infrastructure is inadequate and the existing infrastructure is not maintained properly. This leads to leakages in the water distribution and wastewater collection systems, which results in wastage of precious water resources and shortages. In addition, leakage from wastewater collection systems causes contamination of groundwater resources. In adequate protection of the groundwater resources and their wasteful use in agriculture results in water shortages and water contamination. Trained personnel and adequate capital funding of the water infrastructure depends on the proper collection of user fees, and allocating sufficient budget for this important function of the government.

Although India has passed laws on water cess (user fee), these laws are not properly enforced. As the water infrastructure is aging, the water cess has to be increased for ensuring the funds required for their maintenance. Using smart meters, many cities are now improving the collection of user fees and ensuring sufficient funds for providing water to the populations. The user fees should be structured to promote water conservation and recycling of treated wastewater in communities throughout India.

3.1.3 Storm Water Management

Storm water management is critical to capture the rainwater in our surface water bodies and in groundwater aquifers. Without adequate storm water management in our cities, there is frequent flooding with resultant loss of property and human life. The cycle of droughts and flooding can be prevented by proper forecasting of weather events, and installing green infrastructure which allows the capture of all storm water in man-made and natural water bodies. Rainwater harvesting is becoming a reality in many cities due to government policies that require all homeowners and government entities to provide the facilities for capture of rainwater. A typical rainwater harvesting system that can be used in a building is shown in Figure 3.2.

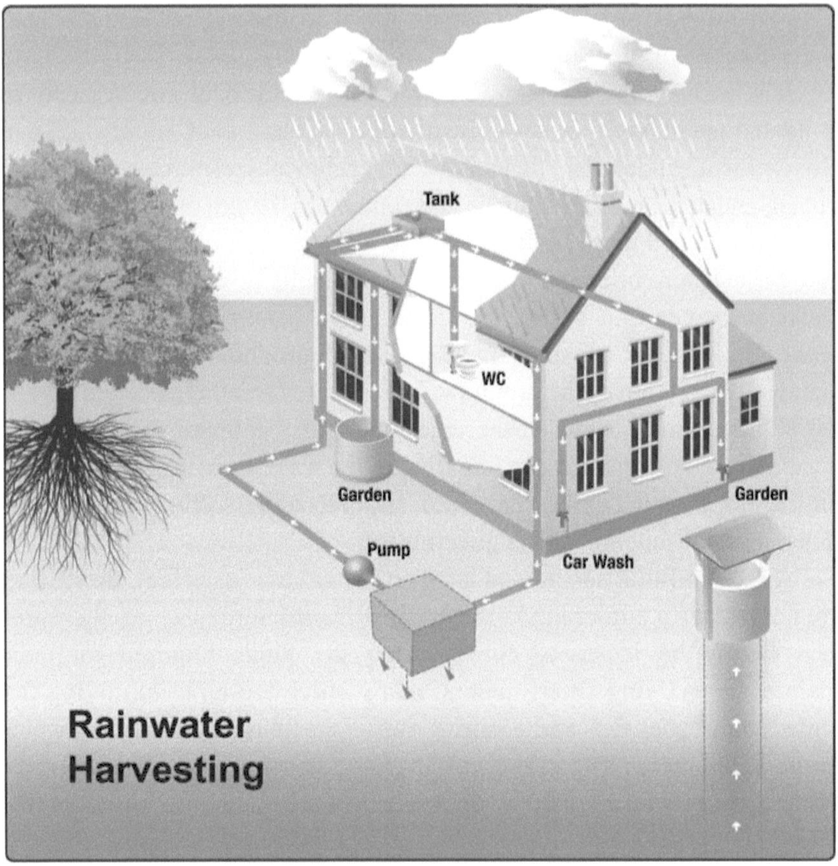

Figure 3.2: ***Typical Rainwater Harvesting system.***
Source: https://byjus.com/biology/rainwater-harvesting/

Other rainwater harvesting/green infrastructure systems include the following:

- Bioswales incorporated into roads and buildings, which allow the rainwater to percolate into the ground;
- Underground and above-ground cisterns to store the rainwater;
- Rain gardens which allow the water to percolate into the ground while enhancing the beauty of the area; and
- Water reservoirs which store the rain and provide for year-round use of the water.

Maintenance of these systems is important for them to function well and provide communities with water while preventing flooding.

3.1.4 Pollution Control

This is a major undertaking with the rapid industrialization of India and other countries. Pollution from industry can be very hazardous since the wastewater from industries contains many hazardous chemicals. Municipal and animal wastes (in rural areas) pollute our water bodies and cause disease through water borne contaminants. Countries have passed environmental laws to control pollution. However, the important part of these laws is rigorous enforcement. India has passed several laws to control water pollution but the lack of enforcement of these laws has resulted in pollution of our rivers, surface water reservoirs and underground aquifers. The environmental laws passed to prevent water pollution include (Bhargava, R.N., et al, 2016,).

- Water (prevention and control of pollution) Act, 1974
- Water (prevention and control of pollution) Cess Act, 1977 and 2003
- Environment (protection) Act, 1986
- Hazardous Wastes (management and handling) Rules, 1989
- Manufacture, Storage and Import of Hazardous Chemicals Rules, 1989 and 2000
- Public Liability Insurance Act, 1991
- Prior Environmental Clearance and Notification, 2006
- National Green Tribunal Act, 2010
- Chemical Accidents (emergency planning, preparedness and Response) Rules, 1996
- Biomedical Wastes (management and handling) Rules, 1998
- Recycled Plastics Manufacture and Usage Rules, 1999
- Fly Ash Notification, 1999
- Municipal Solid Wastes (management and handling) Rules, 2000
- Batteries (management and handling) Rules, 2001
- E-Waste (management and handling) Rules, 2011

All these laws prevent pollution of air, land and water environments, but their lax enforcement is a serious cause of concern to the health of the public. The pollution of these natural resources affects the productivity of the entire population; however, the impact is very significant on the poorer populations who do not have financial resources to counteract the impacts of polluted air, land and water. This leads to many premature deaths in the poorer sections of society and is a cause for environmental justice professionals. Several green tribunals have been set up in India to hear such cases of pollution brought to the court through Public Interest Litigation by well-meaning lawyers. With the intervention of these courts, state and central government departments have taken measures to control pollution.

3.2 Forests

Forests are a storehouse of carbon dioxide, and a wide variety of life forms such as plants, mammals, birds, insects and reptiles. Forests also have abundant microorganisms and fungi, which decompose dead organic matter and enrich the soil. They also absorb a lot of rain, prevent flooding, and are a storehouse of water resources. With the increasing amount of carbon dioxide in our atmosphere from human activities, preserving forests is essential to minimize the impacts of climate change. The Paris Agreement encourages the preservation of forests and allows monetary transfers from rich nations to poor nations to maintain forest cover. Nearly 4 billion hectares of forests cover the earth's surface, which is about 30% of the total land area.

The term forest implies "natural vegetation" of the area, existing for thousands of years and supporting a variety of biodiversity, forming a complex ecosystem. The non-living components of forests are the climate and soil types. Plantations are different in that they are planted species, often of the same type and do not support a variety of natural biodiversity. Depending on the physical, geographic, climatic and ecological factors, there are different types of forests, such as evergreens forest (species having leaves throughout the year) and deciduous forest (species that shed their leaves during the winter months). Each forest type forms a habitat for a specific community of animals that are adapted to it. The subtropical deciduous forests occupy 38.2% of all the forests, and the tropical moist deciduous forests account for 30.3%. Temperate and alpine areas cover about 10% of the forest areas in the Himalayan region.

3.2.1 Forests in India

The India State of Forests Report, 2009, states that the total forest cover is 690,899 square kilometers (sq km), which is 21.02% of the land area. Of this, 83,510 sq km is very dense forest, 297,087 sq km is moderately dense forest, and 269,699 sq km is open forest cover. Scrub land accounts for 42, 525 sq km.

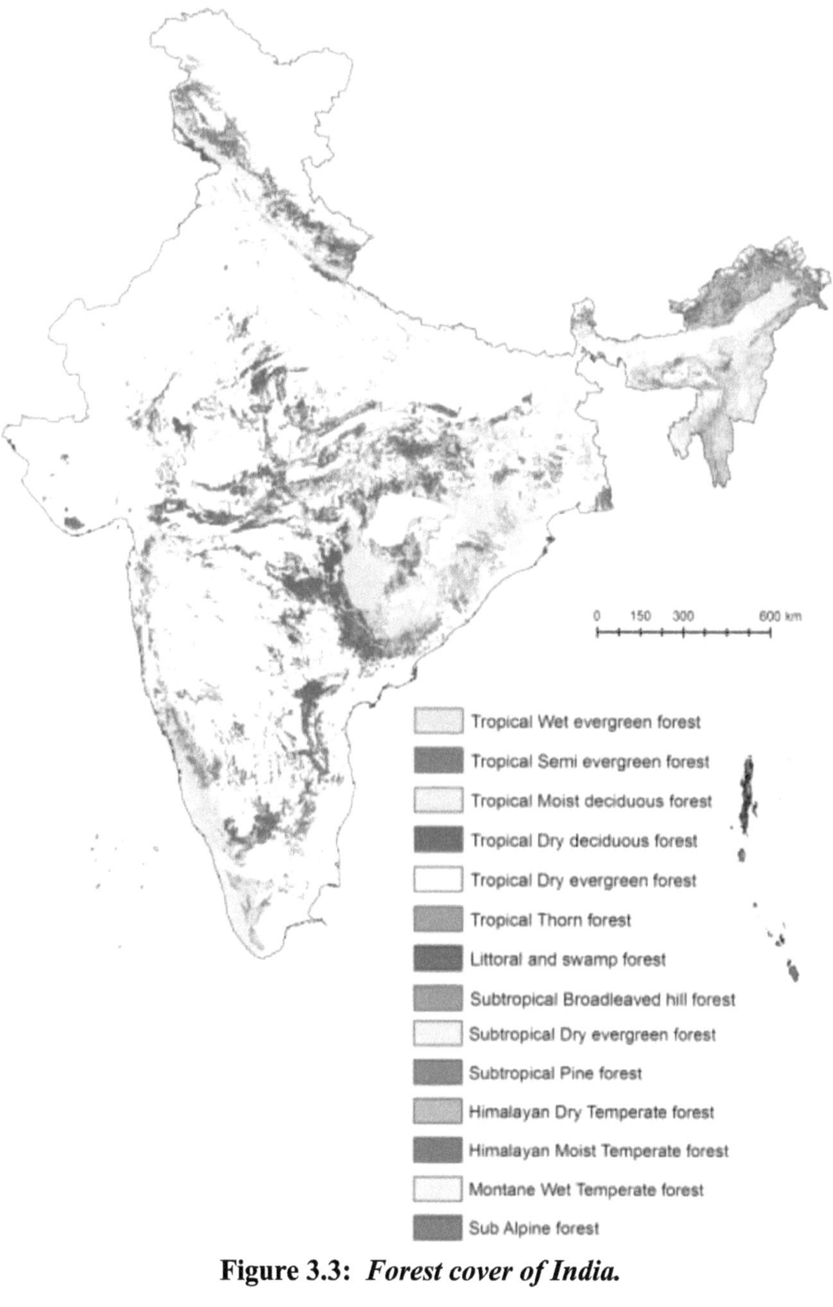

Figure 3.3: ***Forest cover of India.***

Tree cover in India is defined as any group of trees less than 1 ha in size, and having a tree canopy density of less than 10%. The tree cover was 92,767 sq km in 2007. Tree cover is extensive in the states of Maharashtra,

Gujarat, Rajasthan and Uttar Pradesh (Bhargava, R.N, et al, 2016, Ecology and Environment, New Delhi, TERI).

Forest resources worldwide are under attack from developers and others who want to make a short-term profit from this natural resource which belongs to the people. The Amazon Forest in Brazil is huge but constantly, the local people are driven out and big companies move in to log the forest irresponsibly or cut down forest for large scale agriculture. Unless forest resources are replaced with afforestation, the valuable forests are lost forever and have a huge negative impact on climate and we lose the other benefits provided by forests. In India, mining companies in Orissa and other states have been moving into forest reserves and displacing the local people while making huge profits out of the mineral resources under these forests. Mining clearances are given without adequate study of all environmental impacts, and strict provisions for restoring the forests after mining is completed are not put into place.

Sustainable forestry provides livelihoods for the furniture, home building, and many other industry sectors which depend on forest resources. Constitutional provisions empower local governments to use non-wood forest products. India's strategy for forest management focuses on empowerment of community institutions for managing and deriving livelihoods from forests. The Government of India has promulgated a law recognizing the rights of forest dwellers wherein ownership rights are documented and the rights of communities are recognized. These rights include the common use of forests for their livelihood practices conforming to sustainability of forests. The outcome of this law will be known after a few years. Forest officers ensure that the rights of animals and people are balanced and that afforestation is done to sustain the forest for future generations. Several large Tiger Reserves and National Parks are part of the forest management plan of the government.

3.3 Mineral Resources

India is endowed with a lot of mineral resources which are mined for producing energy, construction of infrastructure, and supporting a modern society (Bhargava, R.N, et al, 2016, Ecology and Environment, New Delhi, TERI). Data on Mineral Resources is maintained by the Indian Bureau of Mines. The data is updated annually in the form of the Indian Minerals Yearbook. This contains the mineral prospects, deposits, mines in freehold and leasehold areas, their status, infrastructure, geology, ore characteristics, estimated reserves/resources, and details of mining along with the production data. The data is obtained from various agencies, including the Geological Survey of India, State Directorate General of Mines, and public and private sector mining organizations. The report currently comprises 16,000 deposits, of which 8000 are in freehold areas, 800 are in public sector leaseholds, and the rest are in private leaseholds.

India has large reserves of coal, iron ore, bauxite, chromium, manganese, barite, rare earths and mineral salts. A map of the coal deposits in India is provided in Fig. 3.4

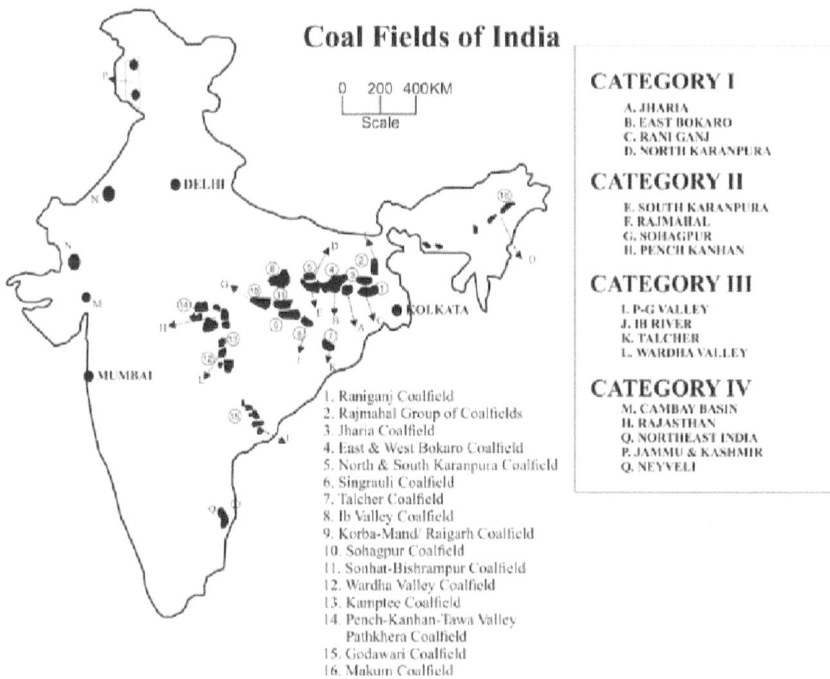

Figure 3.4: ***Major Coal Fields of India.***

There are more than 3700 mines, employing more than 500,000 people. The value of mineral production is about 2.4% of the gross domestic product. Of the total production, fuel minerals account for 64%, metallic minerals account for 15%, non-metallic minerals account for 3% and the rest is construction materials. Mineral and Mining Sector legislation is extensive and includes laws covering the exploration, mining and processing of minerals. These include:

- The Mines and Minerals Act of 1957;
- The Mineral Concession Rules, 1960;

- The Mineral Conservation and Development Rules, 1988;
- The Offshore Areas Mineral Act of 2002;
- The Offshore Areas Mineral Concession Rules of 2006; and
- The Mines and Minerals Act Amendments of 2015.
- Land and Agriculture and Urban Green Spaces

India is an agricultural country with over 60% of the population living in farms. India has become self-sufficient in food supply with the Green Revolution (using hybrid seeds and modern farming) and white revolution (using improved dairy farming and distribution systems).

India has 1.3 million square miles of land, protruding into the Indian Ocean and located between the Bay of Bengal on the east and the Arabian Sea on the West (Bhargava, R.N. et al, 2016, Ecology and Environment, New Delhi, TERI). The types of land include 43% in the plains, 30% in the mountains, and 27% in the plateaus.

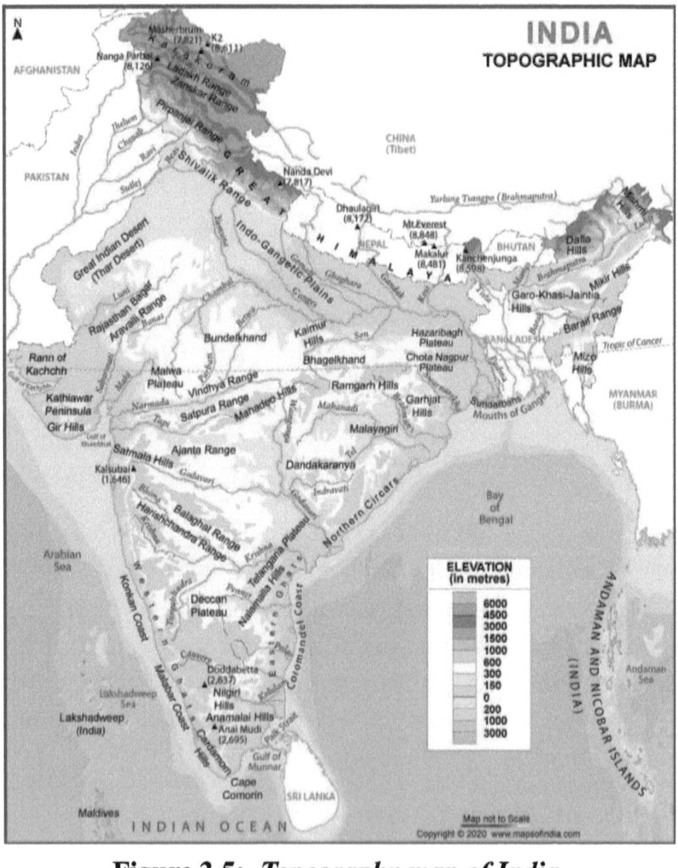

Figure 3.5: ***Topography map of India.***

Increased urbanization and industrialization have put a lot of pressure on agricultural land. In addition, population increases are putting pressure on the land availability for housing in urban areas, and for land availability for infrastructure.

Threats to green space usually are in the form of:

- economic development purposes
- recreational uses (sports, hiking, camping) which creates compaction, litter, erosion & flooding,
- lack of funds for the maintenance of green spaces

The Indian government is constantly improving its efforts in land management, and using waste lands through rehabilitation programs. Reclamation of mined lands is also a priority of the government. Agricultural land occupies 56.8% of the land, and many cities occupy the remaining land. Land resources include many large barren lands in Rajasthan (desert land in the northwest of India), and parts of Leh and Jammu (mountainous areas). A wide variety of crops are grown and with warm weather in most of the country, two to three crops are possible. Wheat, rice, maize, and vegetables are grown, with rice being a major staple in south India and wheat a major staple in north India. Cattle grazing is mainly for dairy and lamb and chicken are the main meats in the country. Many farms are small, and family owned. Some large agricultural producers are consolidating many small farms to produce large commercial crops. Coffee, tea, nuts and spices are the major export crops.

3.5 Air

Clean air is the right of every citizen in India and the world. However, increased urbanization and large population growth in several major cities around the world is deteriorating the air quality in many areas. Many governments have been passing Clean Air Quality legislation from the 1970s to preserve clean areas for citizens. India has passed a Clean Air Quality Act and government agencies in the State and Central Government monitor air quality and publish it daily. The northern part of India, especially the capital New Delhi, suffers from poor air quality in the winter due to climatic conditions and vehicular traffic. Burning of agricultural residues also contributes to smoke and poor air quality. Air Quality Reports are published for all urban centers around the country, and this gives the citizens an idea of the air quality they are breathing. Air quality maps for the country are published regularly, and this is used as a basis to improve the air quality in urban areas.

South India is relatively less populated and enjoys good air quality. The government is devising many methods to improve air quality and they include controlling vehicular traffic and the switching of government and public vehicles to clean natural gas instead of the polluting diesel fuel. This has improved air quality in many major urban areas of India.

3.6 Activities and Exercises

All localities and major cities and nations in the world are facing concerns regarding the depletion of natural resources, destruction of green spaces, and pollution of water, air, and land resources. The following are activities for research and investigation to help students appreciate and understand the challenges.

1. If you live in an urban environment, there is a good chance you are a bit nature deprived and lack first hand experiences and observation of nature. So a first step would be to conduct some research to create a list of what natural areas are available nearby. A person needs to study local habitats so one can really investigate nature in a hands-on manner and if possible see the results of an action. If one simply studies about rainforests or animal extinction on another continent, it is harder to truly feel connected even if you adopt an acre of the rainforest. There is still a disconnect. Do a survey to see what types of resources may be available nearby. Here are some ideas:
   - School Yard Spaces—one can see squirrels, insects, birds, perhaps collect soil, some trees/plants
   - Local community gardens or green roofs
   - Protected natural areas, parks, and nature preserves
   - Local city parks and waterfronts
   - NGO's, universities, or other organizations may offer special programs to take people into nature or may possibly have supplies or materials available for use, microscopes, nets, collection jars, and experts to lead walks and talks

   Once you identify options, go into these local natural areas. Keep a journal, draw, and observe nature, and identify wildlife. Visit during different seasons so you can see changes. Formulate questions and create your own scientific inquiries. You can study such topics as plant structures, insects, erosion, soil composition, rainfall averages, and identification of species.

2. Creating a garden near a home, public place, or school or even on a rooftop can immerse one in all many environmental issues and provide for much learning and appreciation of nature. Seeking to plant with native plants and make a garden bird and wildlife habitat makes the project even more important. This task entails researching bird friendly plants, building and making bird feeders, learning about soil, water, climate conditions that promote growth, and actually growing plants. If one is successful in attracting animals and insects, observation of animals will be easier. See these sources:
   - https://www.nwf.org/Garden-for-Wildlife
   - https://cere-india.org/native-biodiversity-garden/
   - https://www.downtoearth.org.in/interviews/urbanisation/-green-roofs-can-help-cities-adapt-to-climate-change--50850
3. Animals and plants can be potent symbols for a country. For India, answer the following questions, (answer similar questions for your state or community): What is:
   The national animal?
   The national bird?
   The national flower?
   The national emblem (and what animal or plant is part of the emblem)?
   What animals or plants are on the currency - bills and coins both?
4. What national parks for animals or nature sanctuaries or preserves are in your state and all of India? What is their purpose and what problems do they have achieving that purpose?
5. Work with local green spaces to plant, clean up, or inventory wildlife at that green space. Perhaps make it more bird or butterfly friendly. Do a cleanup of some area such as a park or other public open space or of a street or walkway or of a stream. Summarize what you have found by weight or volume or type of material or combination of those. Take pictures and publicize the result, or invite the press to do the same.
6. Inventory, measure, and map all green spaces in your community. Calculate green space available per person and trees per person are other activities. See: www.greenmap.org.
7. Go with a group on field trips to a site of environmental interest. The choices depend on what's available in your area, but could be natural areas and nature preserves, developments, mines, waste disposal sites

or landfills, recycling centers, waste treatment facilities, power generation stations, chemical plants or even 'dumps' where disposal was done illegally or improperly. Before such a trip, have some idea of what to expect and follow all guidelines. Record your observations and ask questions of guides. Formulate ways to establish relationships with these entities and work for change to preserve green space or minimize impact on green spaces where needed.

8. Investigate the web for other examples and resources for green space and habitat preservation and creation. A few interesting ones:
   a. National Arbor Society. Free resources about tree planting. www.arborday.org.
   b. Wildlife Conservation Society. Promotes protection of habitat to preserve tiger species. http://www.wcs.org/where-we-work/asia/india.aspx
   c. Chicago Wilderness Society. Model of restoration in an urban area. www.chicagowilderness.org.
   d. Sustainability focused Organizations: U.S. A https://www.sustainable.org/ and India's Center for Science and the Environment http://www.greenschoolsprogramme.org/about-cse/
   e. Greening School Grounds, Tim Grant and Gail Littlejohn. New Society Press: Toronto, Canada 2001. Available through Green Teacher, P.O. Box 1431, Lewiston, NY 14092. https://greenteacher.com/
   f. Project Wild and Project Learning Tree. Collection of Lesson Plans for teaching about nature. http://www.plt.org and http://www.projectwild.org/
9. *Watershed Preservation and Protection*—Where does your water come from? How is the watershed or source of drinking water being protected from pollution and development What are the costs of building filtration systems and how will this affect the cost of water?
10. *Water Quality*—Environmental agencies must perform tests to ensure that water is free from contaminants such as microbes, inorganic contaminants, pesticides, herbicides, organic chemical contaminants, and radioactive contaminants. How does this work in your locality? How is bottled water produced and tested?
11. *Lead Contamination*-Water can pick up lead from solder, fixtures, and pipes found in the plumbing of some buildings and homes. People who live in many older buildings may have a greater risk. See if you can order a lead testing kit and do an analysis.

12. *Pollution of Waterways, Watersheds, and Ground Aquifers*
    - *Point Sources (easily identifiable sources)*—Chemicals and heavy metals from factory wastes; fertilizers, pesticides and herbicides from farming; animal waste and bi-products from dairy, poultry, hog, cattle farms and production facilities; and a toxic brew of chemicals from city sewers and landfills, as well as leakages from septic tanks, gas storage tanks; are the main sources of pollution. Investigate major threats in your community and efforts to educate and reduce these threats. Start a campaign yourself.
    - *Non-point Pollution sources—pollution from dispersed and hard to identify locations that enter the waterways*—During significant rainfall, sewers overflow, carrying combined household and industrial wastes directly into waterways. Materials discarded on streets, sidewalks can be carried directly into waterways. Paved areas or areas cleared of vegetation create quicker run-off. Individual use of pesticides and fertilizers and improperly disposing of chemicals and cleaning materials, and expired medications all contribute to pollution in our waterways. Investigate major threats in your community and efforts to educate and reduce these threats. Start a campaign yourself.
13. *Controversy on Fluoride Treatment*—Some believe it is not necessary to add fluoride to water. Fluoridation may be worthless, creates discoloration of teeth (dental fluorosis) and is harmful when swallowed. Calcium and good nutrition are more important for preventing tooth decay. Investigate the status of fluoridation in India and issues surrounding this controversy.
14. *Conservation of Water and Drought*—Only 2% of the world's water is drinkable. In many places it comes from underground aquifers that replenish slowly. In certain geographical areas, reservoirs do replenish naturally, but it is disrupted when there is a shortfall of rain or snow. With global weather patterns changing droughts are becoming huge threats. Limiting our use of water through many measures, such as shorter showers, turning off faucets, as well as reducing consumption, changing manufacturing & farming processes, and changing to a vegetarian diet are all water conservation practices. Research these types of practices and see if there are any education efforts in your community to improve water conservation. Start some if not.

You can also calculate how much water you and your families are using both at school and home, search for drips/leaks, and strategize methods to conserve better.

- http://www.waterfootprint.org/?page=files/home or Information on Water Use around the world, water footprint calculator:
- http://www.waterpressures.org/

a. Consider rain water and how it is handled. Many parts of India have too much water during the monsoon season and sometimes not enough part of the year. Water harvesting is a general term that is one way of addressing the problem of uneven distribution of water during the year. The collection could be from roofs into barrels. The water could be directed into rain gardens or a pond. Water will also infiltrate into the ground and maintain the water table so wells don't go dry. Expand on this in ways appropriate to your house or community and part of the country. What water supply problems are a concern in your community and what water harvesting methods are best suited to the climate of your town and area.

15. *Fish/Seafood and Human Consumption*—Toxins in waterways concentrate in seafood flesh and can harm humans when ingested. In addition, overfishing and unsafe fishing practices of many species is destroying the biodiversity, habitat, and balance of our oceans and waterways. Research the status of these issues in India. See sources such as:

    Conservation International https://www.conservation.org/projects/Pages/Ocean-Health-Index.aspx

16. *Sewage Treatment Plants*—Sewage treatment plants are necessary and must follow laws regarding their operation, but humans must be concerned about the gases, sludge, and nitrogen rich treated wastewater released by them. Investigate how your municipality's wastewater is treated and any concerns in your community.

17. *Revitalization/Preservation of Waterways*— Conservation and health of our waterways is crucial to survival of species and recreational enjoyment. Research the efforts to address preservation of rivers, lakes, wetlands, or other waterways in your area. Research ways that these places may be under threat from development or human over use. Studying what's happening with the Ganges River would be a great start.

18. *Access to Waterfront*—Access to waterfront for recreation is limited in many areas of the world. Development must include this option, but is often overlooked. Investigate whether your locality or other cities in India are addressing this through plans for the revitalization of waterfronts.
19. *Water Rights* –Bottled water is big business. Many national corporations are on a quest to purchase rights to aquifers and sell bottled water worldwide, sparking much conflict, depletion of water for local municipalities, and water shortages. Some fear many future wars will be over water as worldwide supply diminishes. Research how this issue is affecting India.
20. *Climate Change and Water*— Scientific evidence indicates that world temperatures are increasing. Many of the effects can be projected but the extent and timing remain somewhat uncertain. This climate change is causing sea levels to rise. Coastal areas across the globe could be severely affected if this trend continues. See what places in India are under threat.
    A. One effect that has been noticed is the melting of glaciers. If the glaciers disappear in the Himalayas, the seasonal flow in many of the major rivers of India will vary even more widely than currently, from very little flow to more extreme flooding. Research to what extent this has already happened. If you live near a river that originates in the Himalayas, ask older people if they have noticed any difference in the flows now compared to decades ago.
    B. If the oceans were to rise as much as one meter, a projection possible if there is substantial melting of the ice caps on Greenland and Antarctica, what areas of India and Bangladesh would be most affected or even inundated. Remember that areas don't have to be normally under water to be affected, as storm surges from a typhoon carry well inland. Also, current coastal areas such as mangroves that can buffer wave action may not survive in deeper water. What are some of the scenarios both for climate change and for the possible effects of the change? See: Clean Air Cool Planet. www.cleanair-coolplanet.org
    C. Ocean acidification due to climate change is an often overlooked aspect of climate change. Research the issue and work to publicize it by writing reports for local newspapers or spreading links to issues on social media. See reports such as: https://unchronicle.un.org/article/climate-change-poses-threat-our-oceans

21. *Environmental Activism*—The World Wildlife Fund, UN, and International Institute of Sustainable Development are global advocacy organizations that work on protecting natural resources. In the United States of America, the Audubon Society, National Resources Defense Council, Greenpeace, and Sierra Club, monitor and work to protect local waterways, wetlands, rivers and oceans and the wildlife that inhabit them. In India, the Center for Science and Environment and Center for Environment Education are two active groups. Research the work of these groups to understand their methodologies and get involved.
22. Mapping/Study of Sources of Pollution in Communities. Students can map locations of producers of toxins and pollutants that could contaminate water, air, or soil in their neighborhood. High traffic areas, factories, waste disposal sites, along with homes of severe asthmatics in their school would be good places to start. Lobbying of government or business entities to change practices could be done to change some of these situations if appropriate.

References

Bhargava, R.N, V. Rajaram, Keith Olson and Lynn Tiede, 2016, Ecology and Environment, New Delhi, TERI.

Chapter 4

Energy and Its Impact on Climate Change

Introduction

Energy is required for carrying out the activities of daily life in both urban and rural areas of the world. The poor in rural areas use wood as an energy source, creating deforestation and significant impacts on climate change. Unless they are given viable alternatives such as solar or biogas energy, they will continue to use wood for cooking and staying warm, which will lead to deforestation. Urban areas use coal and oil for generating electricity, and create significant increases in greenhouse gases. Nuclear energy accounts for 15 to 20% of energy generation in the US and about 80% in France. Countries like India and China are trying to increase energy generation using nuclear power and thus reduce greenhouse gases from the use of fossil fuels. Hydroelectric power is clean but it creates other social and environmental impacts. The only bright spot in the energy generation sector is the use of large solar and wind power projects in many parts of the world. For example, Germany has increased the use of renewable energy and reduced the use of fossil fuels and nuclear energy. The Paris agreement also aims to increase the use of renewable energy in many parts of the world. Innovation is the key to implementing the Paris agreement by many governments around the world.

4.1 Use of Energy – Historical Trends

Let us start by the simple and widely used definition of "energy" as the *capacity to do work.* This definition is deceptively simple and somewhat confusing since work is only one form of energy. The most common are as follows:

(i) *Kinetic energy*: Energy possessed by a system caused by the velocity of molecules; for example, the wind turbines turbine blades and electricity is generated.

(ii) *Potential energy*: Energy possessed by a system due to its elevation; for example, water behind a dam flow down through a pipe and turns a turbine which produces electricity.

(iii) *Chemical energy*: Form of stored energy in a body that is released as heat when burnt; for examples, in fuels such as coal, natural gas, biomass.

(iv) *Nuclear energy*: The energy released when the nucleus of certain material (such as uranium) is split.

(v) *Solar energy:* Electromagnetic radiation from the sun which can be converted into thermal energy or electricity.

We will note that electricity does not occur naturally but is obtained by conversion from the above which are referred to as *primary sources of energy*. The commonly used units of energy are Joules (J) in SI units and British Thermal Units (Btu) in IP units. These units are very small and so prefixes are used. Table 4.1 assembles some of the commonly used prefixes for energy quantities.

Table 4.1: *Commonly used energy prefixes for energy quantities*

Prefix	**Value**	**Name**
Mega	10^6	Million
Kilo	10^3	Thousand
Centi	10^{-2}	Hundredth
Milli	10^{-3}	Thousandth

It is important to distinguish between energy and power. Power is defined as energy per unit time. The most common unit is Watts (W) defined as Joules/second. Another commonly used unit of power is the Horse-power (HP equal to 746 W).

Example: A crane lifts a block of concrete weighing 1 tonne to a height of 30 m. How much energy has been expended?

1 tonne = 1000 kg

Potential energy = $1000\text{kg} \times 9.81\text{m/s}^2 \times 30\text{m} = 294300 \text{ J} = 294.3 \text{ kJ}$

The above is accomplished in 10 seconds. What is the power of the crane?

Power = Energy/time = 294.3/10 = 29.43 kW

Human beings burn energy to maintain their body temperature. Thus, they release energy called basic sustenance level. This amount of energy is around 100 W, i.e., 100 J/sec every second of every day. For the sake of comparison, the maximum human power output over one minute is 2000 W. The energy consumed by prehistoric man for hunting and food gathering is about 125 W. This gradually increased to about 500 W during the industrial age about 200 years back, and then there was a huge increase in modern times of around 10,000 W per person. This huge per capita increase in energy use is attributed to be the driver of modern economic prosperity. Thus, energy is a basic and critical factor to sustain our modern way of life.

This large increase in energy need is partly because of the large increase in human population, and also due to a higher standard of living. Figure 4.1 illustrates the relative increase in population (in billions) and energy consumption in Q btu/yr. (or quadrillion Btu = EJ/yr.). We notice that energy consumption is increasing at a faster rate than its population. Another important aspect to note is that the developing countries will be the reason for the large increase in the next 80 years due to both population increase and an increase in their standard of living.

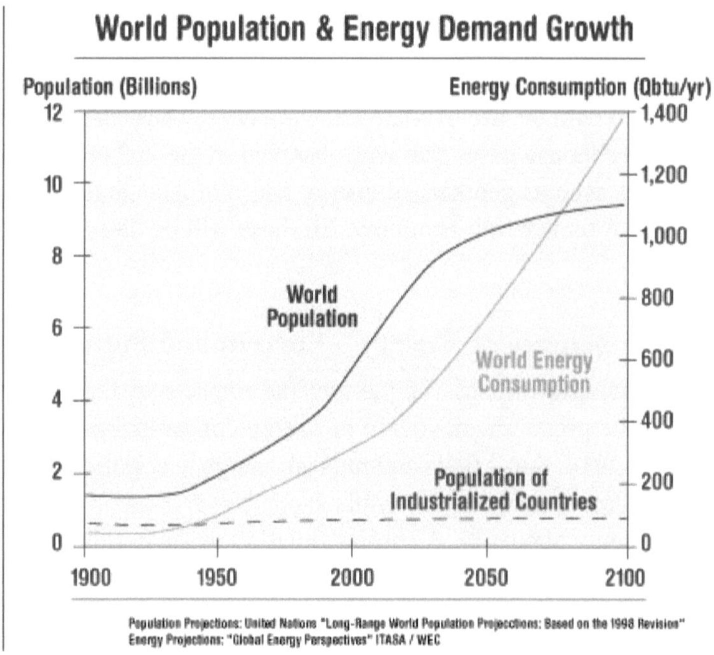

Figure 4.1: ***Past and future population and energy consumption trends.***

4.2 Forms of Energy in the 21st Century

With the dawn of the industrial revolution in the 1700s, the use of coal and wood accelerated and thus exacerbated the greenhouse gases in the atmosphere. Since then, the discovery of huge oil reserves in the US, Canada, Saudi Arabia and other parts of the world led most industrialized countries to use oil for electricity generation and transportation. The discovery of nuclear energy in the 20th century slowed down the use of coal and oil for electricity generation; however, the enormous costs, safety issues and environmental concerns have slowed the growth of nuclear energy. Huge advances in the design of solar cells and wind turbines have accelerated the use of renewable energy; however, the storage and transmission of renewable energy pose several challenges which are being addressed in the 21st century.

The advances in the 21st century have been amazing, especially with the dawn of the internet age and the computer revolution. The means of monitoring our energy usage, and improving efficiency of energy usage have been phenomenal. This has saved a lot of electricity generation since energy efficiency is the most inexpensive way of reducing our energy usage, and once the improvements are implemented, they save energy year after year. Discovery of new technologies to tap the large reserves of natural gas trapped in tight shale formations has increased the use of clean natural gas in transportation, heating our homes, and electricity generation. This, combined with the use of solar and wind energy has curbed the steep increase in greenhouse gases that was observed in the 20th century. Other energy sources such as geothermal energy and biofuels are also reducing the use of fossil fuels in our economy. All these will be discussed later in the Chapter.

4.3 Primary Sources of Energy – Current and Future Trends

Despite increasing awareness of the adverse impacts of fossil fuel use, they are still the predominant source of energy consumption as shown in Figure 4.2. In 2014, fossil fuels and nuclear energy accounted for 90% of the global energy consumption, while hydropower was 7% and solar-wind-bio was only about 3%. A similar situation exists in India as well as shown in Figure 4.3, where renewables make up only 3% of the primary energy use.

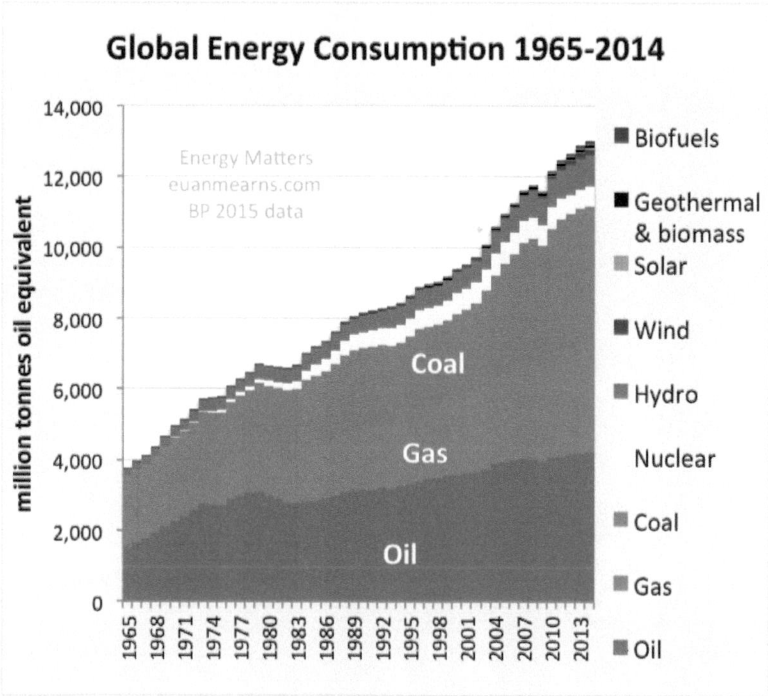

Figure 4.2: ***Predominant sources of energy consumption.***

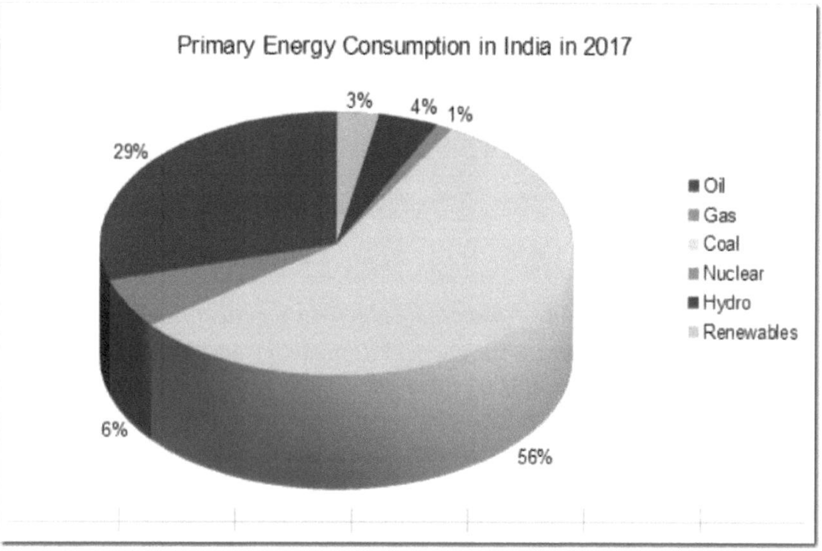

Figure 4.3: ***Primary energy consumption in India in 2017.***

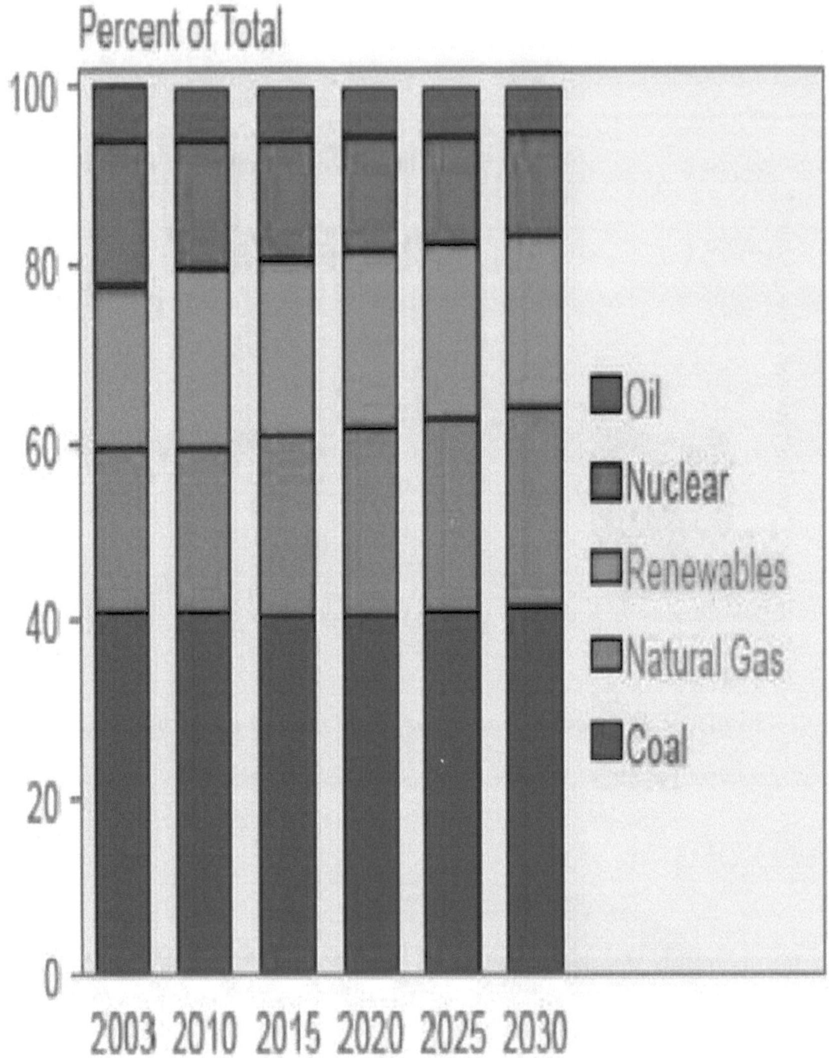

Figure 4.4a: ***Sources of electricity generation.***

Sources: 2003: Derived from Energy infromation Administration (EIA) International Energy Annual 2003 (May-July 2005), website www.eia.doe.gov/iea/. 2010-2030: EIA, system for the Analysis of Global Energy Markets (2006).

Note: Fuel Shares may not add to 100 percent due to independent rounding.

A large portion of the fossil fuels are used for electricity generation and the rest for transportation. From Fig. 4.4a, we notice that renewables (which include hydropower, biomass, solar and wind) account for less than 20% of the electricity generated, and this trend is likely to continue for the next decade into 2030. The great predominance of fossil fuels in the Indian society is obvious from Figure 4.4b where we notice that 59% of electricity generation is coal-based, with an additional 10% from natural gas and oil.

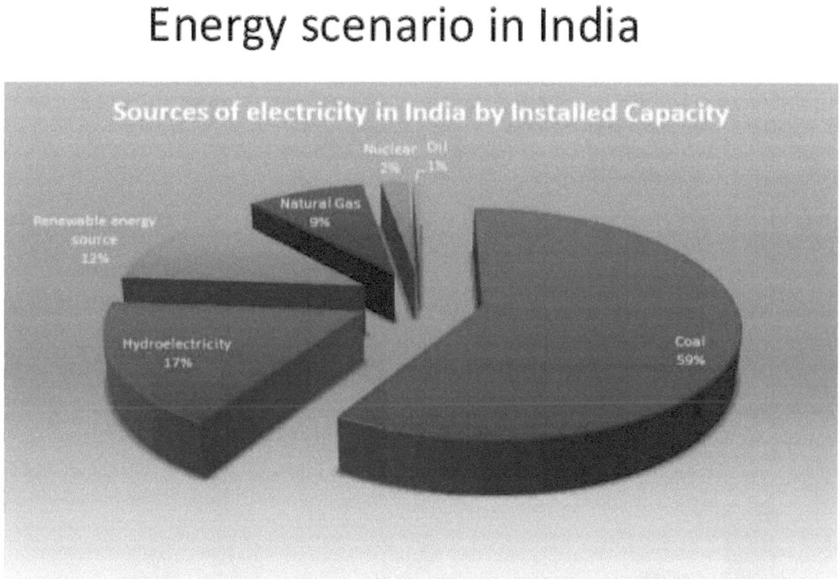

Figure 4.4b: ***Sources of electricity in India by installed capacity.***

Where is this electricity consumed? The various end-use sectors for electricity consumption are shown in Fig.5. We note that industry accounts for 40% of the electricity generated, with about 33% in commercial buildings and in residences. It is worth noting that for the United States, commercial and residential buildings account for over 70% of the electricity generated. This trend seems to be common in developed countries in general. However, the growth rate of electricity use in buildings is projected to increase greatly in the future as shown in Fig. 6. While the developed countries are shown to have very low growth rates (only 0.1% for the United States), that of India is 2.7% which is even greater than China at 2.1%. So, finding ways to reduce electricity use in buildings will assume major importance in the future. The best way to reduce electricity use while not compromising delivered services, is to improve appliance efficiency and implement aggressive energy conservation measures.

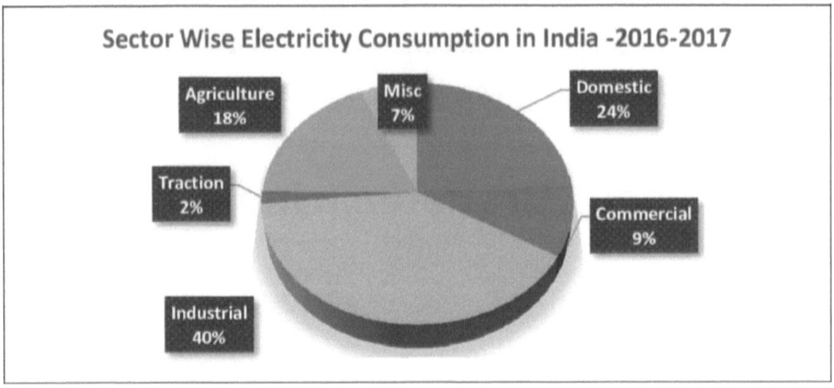

Figure 4.5: ***Sector wise electricity consumption in India 2016-2017.***

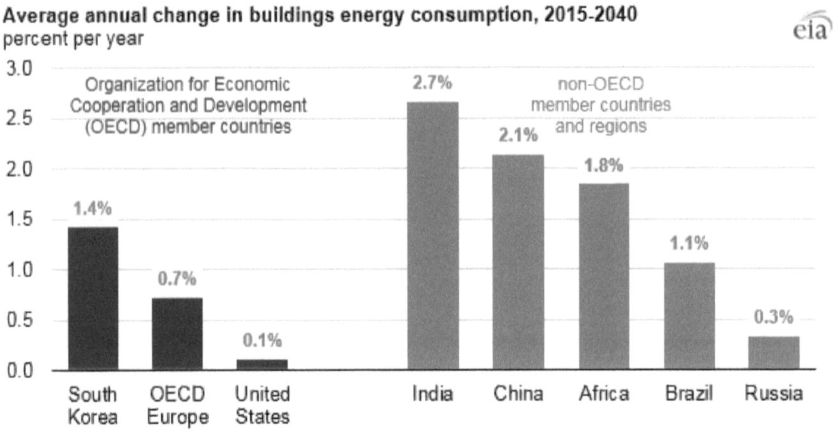

Figure 4.6: ***Average Annual change in buildings energy consumption 2015-2040.***

4.4 Fossil Energy and Greenhouse gases

The rapid increase in the use of fossil fuels in the 19th and 20th centuries coincides with the rapid increase of greenhouse gases in our atmosphere, and this in turn has exacerbated the natural disasters related to climate change. Coal is being widely used in China and India, and despite efforts to introduce renewable energy in the energy mix, these two countries will continue to use a large amount of coal to meet their growing energy demands. Severe droughts, flooding, hurricanes and typhoons, and hotter summers have had a disastrous impact on our economies and on the human and animal populations around the world. The most vulnerable population impacted by these natural phenomena are the poor, and they are not enjoying the benefits of fossil fuel use that go to middle class and higher income populations. Technological innovation is the only way to minimize the use of fossil fuels and thus minimize their impact on greenhouse gases. Citizen involvement in the creation of public policy that encourages alternatives to fossil fuels is the only way to bring many technologies to the marketplace.

The industrial revolution, the two World Wars, and other economic/ social changes around the world have impacted the growth of fossil fuel use. The dawn of the jet age has made worldwide travel easy but in turn it has significantly increased the use of jet fuel and the emission of greenhouse gases. Just as the green revolution increased the availability of food around the world, technological innovation has allowed us to replace coal and oil with natural gas, nuclear energy, solar and wind power and biofuels. The Paris agreement is a significant step for the world taking greenhouse gas induced climate change seriously and focusing on technological innovation and public policy to reduce the use of fossil fuels.

The coal and oil industries have spent a lot of money on research to improve combustion technologies and capture the emissions from power plants. Technological innovations in the efficient and safe use of fossil fuels in transportation and electricity generation have made the emissions from vehicles less harmful to the people and improved the miles per gallon for fuel used in motor vehicles and airplanes. Efficient use of electricity in our buildings and homes has reduced the greenhouse gas emissions from fossil-fired power plants.

4.5 Nuclear Energy

The first use of atomic power was in World War II when Japan was bombed by the United States. Since then, much work has been done on the peaceful use of atomic power and many power plants have been built using atomic power. The safety considerations and enormous costs of building nuclear power plants has slowed the growth of nuclear energy. In addition, the high cost of disposal of nuclear waste is discouraging the use of nuclear power. However, since nuclear power plants do not emit greenhouse gases, the use of nuclear energy is increasing in India and other countries like France and China.

The fear of nuclear radiation's long-term health impacts (as documented from the bombing of Hiroshima and Nagasaki in 1945) has almost stopped the development of nuclear power plants around the world. The extraordinary safety measures that need to be taken in nuclear power plants significantly impacts the cost of building and running nuclear power plants. The nuclear waste resulting from the power plants remains dangerous for over 10,000 year and so there is no cost-effective way for permanently disposing the nuclear waste. There are several million tons of nuclear waste accumulating around power plants around the world, and this poses a major danger to future generations. Hence, the future of nuclear power is bleak, and more emphasis is being paid on renewable energy plants worldwide. See Chapter 12 for the Fukushima Disaster case study, which includes a discussion of possible radiation from nuclear power plants.

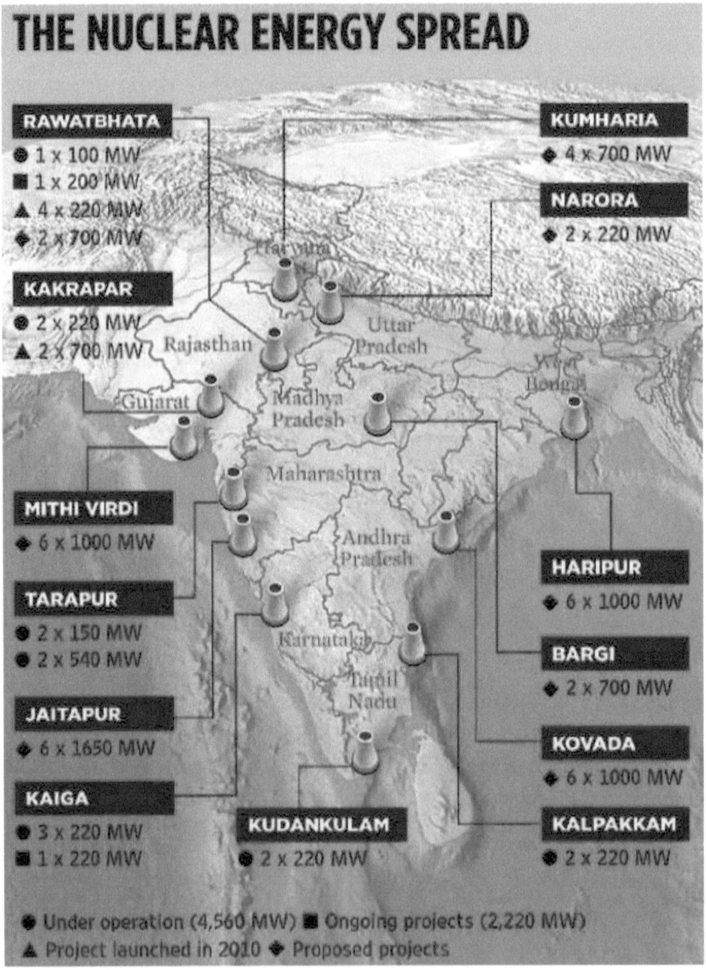

Figure 4.7: ***Nuclear power plants in India.***

4.6 Adverse Impacts of Fossil Fuels and Nuclear Energy

- Unsustainable patterns of energy use
- Carbon-based fossil fuels cause global warming, climate change, and expected impacts on sea level rise, extreme weather, human and ecological health risks

  There are a number of costs incurred due to the widespread use of fossil fuels. These are called "hidden costs or externalities". These are costs that are not properly reflected in what the consumer pays in electricity or gas; nor are they paid by the companies that produce or sell energy- society as a whole pays for them or accepts the consequences.

  Examples are:
  - Human health problems caused by air pollution from burning fossil fuels
  - Damage to land/water from coal mining and from oil drilling
  - Health effects on coal miners (black lung disease)
  - Environmental degradation caused by acid rain and water pollution
  - Global warming and its consequences
  - Inequitable social justice and hardship due to scarcity
  - National security costs (such as protecting foreign sources of oil)

4.7 Renewable Energy -Challenges and Progress

Renewable energy production and use has been rapidly increasing as the cost of solar and wind technology has reduced significantly. In addition, a lot of research has been taking place on biofuels and various feedstocks other than corn for producing biofuels. The main drawback of renewable energy technologies is the storage of power; however, a lot of innovation is happening in battery technology, which has applications in energy storage and electric vehicles. The main types of renewable energy are:

- Solar energy, both for electricity generation and hot water heating;
- Wind energy, both onshore and offshore;
- Biofuels, using corn, Jatropha (in India), organic solid waste, and other feedstock; and
- Hydroelectricity, both small scale and large scale.

Solar energy has made the most strides in the last 20 years, and residential solar energy applications are finding widespread use. The co-author Dr. Rajaram has installed one in his house and the energy savings are considerable. His electric bills have reduced by 75%, and the capital cost of the system will be recovered in 5 years with the help of State and Federal government incentives. Community solar is being adopted in many communities to provide energy to public institutions such as schools and government offices. In India and China, many rural communities are off-grid using only solar energy for street lighting and lighting a few bulbs in each home. Large scale solar farms are used by industry and by utilities for producing solar energy for industrial and community use.

Figure 4.8: ***Solar panels in India.***

Government incentives in India are allowing the rapid adoption of solar energy, and as the usage increases, the cost of solar panels is decreasing due to economies of scale and also rapid innovation. In China, rooftop water heating units are widely adopted by large communities for serving their energy needs. The Paris Agreement is further facilitating the spread of solar energy use around the world.

Wind energy efficiency is improving with accurate knowledge of wind speeds at different elevations above ground and the design of better aerodynamic foils for use on towers.

Figure 4.9: ***Windmill Farms.***

Denmark and other European countries are making great strides in offshore wind farms to supplement those on the ground. The west coast of India and many other regions within India are seeing a resurgence of wind power as the cost of wind energy production is decreasing. Improvement in energy storage systems is also helping the wind energy sector expand worldwide.

Biofuels are helping reduce the consumption of fossil fuels in the transportation industry. India has been building biogas plants in rural areas using animal waste and waste biomass. This has been the only source of energy in many parts of India for several decades.

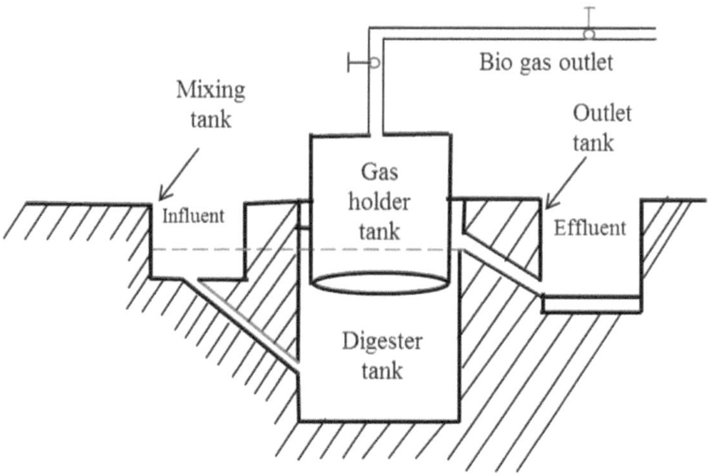

Figure 4.10: ***Biogas Plant.***

India has adopted a National Biomethanation Plan a few years ago, and this has promoted the use of biofuels in all parts of India. The National Policy of not letting any organic food or farm waste go to landfills is promoting the use of community biogas units using the organic food waste in the community (including animal waste). Large scale research is underway worldwide to use alternate feedstocks for biofuel production. This will save the corn for humans and animals, and all kinds of grasses and other organic farm waste can be converted to biofuels. Jatropha is a plant that is grown in many parts of India and Indian companies have made major strides in producing biodiesel from this plant.

Hydroelectric energy was used widely around the world in the 20th century to harness the energy in river waters. However, the cost of large hydroelectric projects and their environmental impact on aquatic life has reduced the use of large-scale hydroelectric plants. However, small scale hydroelectric plants for community use are being implemented in many parts of India and around the world. Meticulous planning is required to study the costs and benefits/harms from these plants before undertaking such projects.

Other forms of renewable energy such as Geothermal (widely used in Iceland) and Wave energy projects are very much site-dependent, and should be explored when the right site conditions are available. Any form of renewable energy that reduces the use of fossil fuels is beneficial to the planet and should be studied from the policy, cost and benefit standpoints.

4.7.1 Wood Energy

Wood has been used as an energy source from time immemorial, and is also being used today in poor, rural communities around the world. The acute awareness that forests and trees are important to counteract the greenhouse gases we emit is leading to afforestation in many areas of the world and reducing deforestation in rural communities. Rural communities in India have been encouraged to use biogas from the animal and other available wastes so that they are discouraged from going to forested areas and removing wood for fuel. In addition, simple solar electrification schemes in rural communities are being undertaken by civil society organizations and government and these are helping with the minimization of wood as fuel.

4.7.2 Biogas

A simple project started by the Engineers Without Borders –Hyderabad, India, chapter uses briquetting of waste biomass to provide fuel to rural communities. Gautam Yadama, author *of Fires, Fuel & the Fate of 3 Billion:*

The State of the Energy Impoverished, has studied the issue greatly linking poverty, poor health, premature deaths in children, loss of opportunity for women and girls for those who still rely on foraging for fuel to chop, collect, and carry in order to burn in open fires that leads to excessive smoke inhalation. It is his belief that switching to clean biogas systems using cattle dung can start to solve all of these issues, and a critical part of the success is training and empowering women to understand the technology and how to fix and maintain it. Better and more research into the most effective systems is also an ongoing need to ensure that people use them effectively and consistently (Hoffner). Groups such as The Fair Climate Fund based in Holland are engaging people and local development groups in India in small scale projects to bring both better cook stoves to regions such as Raichur and to build a village level anaerobic gas plant in ChikBallapur (Fair Climate Fund). These types of projects to connect funders and new technological innovations with those that need cleaner methods of energy production will ultimately lead to global reductions in greenhouse gases.

4.8 Innovation

The pace of innovation in renewable and other forms of clean energy is accelerating, and many venture capitalists are showing a lot of interest in funding such projects. Governments around the world are also putting a lot of resources in innovation for clean energy, and this is getting more impetus from the Paris Climate Agreement. Innovations in monitoring energy usage, and new ways of innovative transportation that do not rely on fossil fuels or use them more efficiently are helping reduce greenhouse gases in all parts of the world. China is leading in innovation in the area of solar energy and India is taking a lead in the area of biomethanation for converting organic wastes into fuels. A breakthrough in the economic conversion of waste biomass from agriculture and other abundant biomass available on the planet to clean energy can have a major impact in the reduction of greenhouse gases.

In addition to innovation, changing the attitude of people in the areas of energy conservation and use of renewable energy will have a major impact on the reduction of greenhouse gases in the energy industry. Young people around the world are being made aware of technological innovation by the world-wide adoption of the internet in education. Universities are creating new programs to educate the students on the responsible use of energy and sustainable living.

4.8.1 Solar and Energy Generating Roof Tiles, Batteries, & Pavement

U.S. based Tesla has been getting a lot of attention for its sleek roof tiles and battery for on location generation and storage. This is a groundbreaking technology that, while expensive at the time of print, is a critical breakthrough for making solar more saleable to every individual. It is thinking and design that is aimed at providing a way to integrate solar into regular architectural materials and also provide more affordable storage of energy where it is generated so that there is less loss, and so that a person who wants to build a solar powered home does not have to rely on far away storage or integration into a corporate power company's grid which may or may not pay the consumer a fair amount for the energy being generated and not used by that consumer. Tesla's goal to create electric cars that could be charged with the solar system integrated into the home is truly a futuristic vision that is closer to becoming reality.

Another interesting innovation has been the development of solar roads. Innovators in China, France, the United Kingdom, and the U.S. have been working on this technology and starting to pilot some prototypes. Solar Roadways in the U.S. received donations from independent donors and a grant from the U.S. Department of Transportation to test a solar road in a parking lot. They believe replacing all US paved roads with solar roads could produce over three times the electricity the entire country uses. UK based Pavegen is developing piezoelectrics which convert the pressure and vibrations from car and foot traffic into usable energy. They have installed one in Washington D.C. which allows pedestrian foot traffic to generate LED street lamps overhead. California has invested in several similar projects on highways to see if such technology can lead to power for homes. According to Green America's Tracy Fernandez Rysavy, "if successful, piezoelectrics could be a real win for the Earth and its climate" (Fernandez-Rysavy).

4.8.2 Algae as a source of Biodiesel fuel

Additionally, researchers in biogas generation are exploring cultivation of algae for its biodiesel fuel. The idea is that if hurdles can be overcome and solved, algae can be farmed at much more cost effective levels than other bioenergy feedstocks and prove a way to provide a renewable energy source for cars, planes, trucks, and trains that normally would use diesel fuel. Developing more technology that runs on this type of fuel could then help shift a transition to algae based fuels once the production on a large scale is developed. Another benefit of this technology is that algae takes a lot of CO2 to grow and develop so it can also be a technology that absorbs existing CO2 from the atmosphere.

4.9 Growth Models for Energy and Implications

Mathematical models are used to predict how certain quantities will increase or decrease over time. These are referred to **as growth-decay curves**. Common quantities could be money in a bank earning compound interest, the population growth of a country or the energy use growth of a country. There are two growth models of basic interest:

- Linear growth wherein the initial amount grows by a fixed amount every time step. Say, we have $1000 in a bank which gives 0% interest and we deposit $100 every year. The money in the bank would grow by $100 each year. At the start of Year 2, the amount would be $1100, start of year 3 it would be $ 1200 and so on. So, the amount grows by a fixed amount each year.
- Exponential model is one where the growth depends on the amount itself. Consider the situation where you deposit $1000 in a bank which gives you 7% interest and you let the interest compound. Then at the start of year 2, the amount will be $1000*(1+0.07) = 1070. At the the start of year 3, it would be 1070*(1+0.07) = 1144.9 and so on. So we notice that the annual increments are increasing: from $70 to $74.9 and so. In fact, at the end of 10 years, the initial amount would be doubled (Table 4.2). The time taken for this to occur is the doubling time. Thus, in this situation, the doubling time is equal to 10 years.

The general formula for doubling time is equal to (0.7/annual rate). Thus if the annual rate was 3.5%, the doubling time could be 20 years. This concept of doubling time is very useful. For example, consider the annual growth rate of the building energy use in India 2.7% (see Figure6). The doubling time would be (0.7/0.027) = 26 years. That means that the energy use would double in 26 years would quadruple in 52 years, and so on.

Growth Rate (%/yr)	Doubling Time
1	70
2	35
5	14
10	7

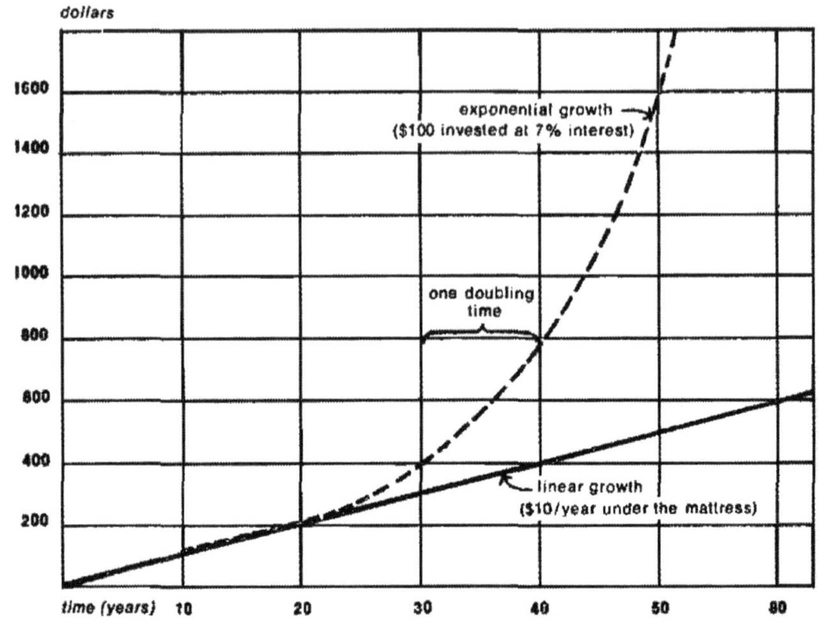

If a miser hides $10 each year under his mattress, his savings will grow linearly, as shown by the lower curve. If, after 10 years, he invests his $100 at 7 percent interest, that $100 will grow exponentiallv, with a doubling time of 10 years.

Table 4.2: ***Doubling time for exponential growth.***

4.10 Activities and Exercises

1. Research current alternative renewable energies being used where you live and create tables to list pros and cons of each source. Many experts believe that the future will need to be a mix of sources of energy, which is different than our current fossil fuel based economy. Understanding resources available in one's local and what types of alternative renewable energy systems will best work in one's environment is a critical part of shifting to sustainable, carbon reducing energy systems.
2. Research innovations in new technologies for energy production like solar roads, tiles, and algae. Use sources such as:
   (a) https://www.energy.gov/eere/videos/energy-101-algae-fuel
   (b) http://news.mit.edu/2013/innovation-in-renewable-energy-technologies-booming-1010 (c)https://www.forbes.com/sites/jamesellsmoor/2018/12/30/6-renewable-energy-trends-to-watch-in-2019/#321dfdd24a1f, 4)https://www.treehugger.com/

renewable-energy/9-energy-innovations-make-future-brighter.html. Brainstorm and create a hypothetical source of energy or a more energy efficient technology using these ideas as inspiration.

3. Along with new renewable and cleaner sources of energy, energy conservation must go hand in hand. Governments understand that broad policies to promote energy efficiency, especially with buildings as well as via individual corporate and consumer habits is critical. Research ways that your country or locality is promoting such efforts and identify how you can develop better practices or advocate for such policies. Some resources can be found at:
   - The Bureau of Energy Efficiency https://beeindia.gov.in/sites/default/files/Download%20Tips%20for%20Energy%20Conservation%20for%20Domestic.pdf
   - National Resources Defense Council https://www.nrdc.org/experts/peter-lehner/indias-next-big-energy-source-energy-efficient-buildings
   - Shakti Foundation: https://shaktifoundation.in/work/energy-efficiency/energy-efficient-buildings/
4. Research the work done by the Ministry of New and Renewable Energy (MNRE), and write a 5-page report. How much of the technologies developed by MNRE are utilized by industry and individual citizens?
5. Study the wind energy usage in Gujarat and how much of the State's energy use is supplied by wind energy? Draw a graph showing the wind energy usage in the different states in 2020. Research the types of energy usage in your city, and how has it changed since 2010? If you were able to design an energy policy for your city, how much renewable energy goal would you have?
6. How do the poor people in a rural area close to your city get their energy? Do they have access to the electric grid? How much wood is used in that rural area? Write a 2-page report about your findings.
7. Where is most of the hydroelectricity produced in India? How much of the country's energy comes from hydroelectricity? Make a list of the major hydroelectric projects in the country, and write a short report on hydroelectricity production and usage.
8. Indian Oil, the largest oil producer and distributor in India, has done a lot of work on biofuel feasibility for cars, trucks and planes. Research

the work done by Indian Oil and other oil companies and write a 5-page report.

9. Mahatma Gandhi urged Indians to use biogas using the animal wastes produced in rural areas. How much biogas is produced from animal waste in rural India? Write a short report on biogas production in the different states, and is this replacing the use of wood for fuel?
10. Natural gas is used for transportation and for cooking all over India. From where does India get its natural gas supply? Are most city homes supplied with piped natural gas? How much gas is sold in metal containers in your city? Do you have natural gas in your home and who is the supplier of the gas?
11. Where are the uranium mines in india? Do they supply the fuel needs of all the nuclear power plants in India or do we buy fuel from other countries? How much of the country's energy is supplied by nuclear energy? Write a report on the history of nuclear power in India.
12. The government is promoting the use of solar energy. Study the National Climate Action Plan published by the Ministry of Environment, Forests and Climate Change, and write a short report on how the government plans to reduce the impact of fossil fuels on Climate Change. How much solar energy is produced in the various states of India? Draw a graph to depict the use of solar energy by State.
13. How much of India's energy comes from coal-fired power plants? Draw a map of India showing the coal-fired power plants. What is being done to capture the carbon dioxide emissions from coal-fired power plants?
14. Does your city produce solar energy? Analyze the energy use in your city, and draw a pie chart showing the various energy sources that supply the energy needs of your city? Is India developing battery technology to store the solar energy produced during the day so it can be used at other times? If so, describe a solar farm project in the country.

References

- Hoffner, Eric. "The Burning Question." World Ark. The Heifer Project, Spring 2015. https://media.heifer.org/world-ark/2015/World_Ark_2015_Spring.pdf

- "India Clean Cooking With Biogas." Fair Climate Fund. https://www.fairclimatefund.nl/en/projects/india-clean-cooking-with-biogas. Accessed 9 July 2019.
- Rysavy, Tracy Fernandez. "Smart Tech for a Greener Life." Green America, Spring 2018. https://www.greenamerica.org/new-green-tech-promises-and-pitfalls/smart-tech-greener- life. Accessed 9 July 2019.
- Solar Roof. Tesla. https://www.tesla.com/solarroof. Accessed 9 July 2019.
- U.S. Department of Energy Office of Energy Efficiency and Renewable Energy. "Algal Biofuels." https://www.energy.gov/eere/bioenergy/algal-biofuels. Accessed 9 July 2019.
- Yadama, Gautam. Fires, Fuel & the Fate of 3 Billion: The State of the Energy Impoverished New York: Oxford University Press, 2012.
- ITASA/WEC, 1998, Global Energy Perspectives.
- United Nations, 1998, Long-Range World Population Projection, Based on 1998 revisions.
- www.euanmearns.com.,Energy Matters, BP 2015 data.
- Energy International Administration, 2003, International Energy Annual Data
- www.eia.doe.gov/iea/2010-2030.
- www.world-nuclear.org/information-library/country-profiles/countries-g-n/india.aspx
- www.researchgate.net/figure/Simple-schematic-of-a-biogas-plant_fig3_318503013
- Figure 4.7 Courtesy: (https://www.world-nuclear.org/information-library/country-profiles/countries-g-n/india.aspx)
- Figure 4.10 Courtesy: (https://www.researchgate.net/figure/Simple-schematic-of-a-biogas-plant_fig3_318503013)

Chapter 5

Biodiversity

Introduction

Biodiversity is given to us by mother Earth but humans have been destroying biodiversity by careless development of all kinds of lands to satisfy human greed. In the village surrounded by a forest, the vacations I spent during Indian summers had become a symbolism of mangoes, cuckoos and parrots. During daytime, we played under the lush Banyan tree that was a favorite abode to a variety of creatures while at night, we listened to stories from grandmother with all the cousins sitting in a circle on the roof-top.

It was an extreme delight to watch the diversity of life forms being nourished on the Indian banyan, "*Ficus benghalensis*" (strangler fig) including sparrows, squirrels, parrots, monkeys, numerous insects and bats.

Figure 5.1: ***Strangler Fig (Ficus benghalensis).***

5.1 Biodiversity

Biodiversity or biological diversity refers to the variability of all living organisms that exist within or among organisms at different hierarchical levels in a particular habitat. It is the sum total of genetic biodiversity, species biodiversity and ecosystem biodiversity.

5.1.1 Genetic diversity

It is the variation in the genes among the individuals of a species in a population. It includes the allelic, genetic or chromosomal variation. Genetic diversity is the total number of genetic characteristics in the genetic makeup of a species. It is distinguished from genetic variability which describes the tendency of genetic characteristics to vary. Genetic diversity serves as a way for populations to adapt to changing environments. Genetic variation can be caused by mutation, random mating, random fertilization, and recombination between homologous chromosomes during meiosis, which reshuffles alleles with an organism's offspring.

Genetic diversity examples include different breeds of dogs, and different varieties of rose flower, wheat, etc. There are more than 50,000 varieties of rice and more than a thousand varieties of mangoes found in India.

5.1.2 Species Diversity

It refers to the variety of different species in a region where the role of every species is important. Species diversity is defined as the number of species (species richness) and the evenness of each species living in a particular area. The number of species per unit area in a given location is called species richness. While the number of individuals in each species represent the species evenness.

Higher species richness and evenness of a community contribute to higher species diversity of that area.

5.1.3 Ecosystem Diversity

Ecosystems like forests, grasslands, ponds include the biotic (living) components such as plants and animals and abiotic (non-living) components like water, soil, temperature, nutrients content of soil etc. of a region and the interactions among them. The functional aspects of the ecosystem involve the energy flow, nutrient recycling and decomposition. This is discussed in Chapter 6.

5.2 Biodiversity in India

India is a megadiverse country. Just 17 of the world's 190 or so countries contain 70 percent of its biodiversity, earning it the title "megadiverse." India has 2.4% of the land area, accounting for 7 to 8% of the species of the world, including about 91,000 species of animals and 45,500 species of plants. This has been documented in its ten bio-geographic regions. Of these 12.6% are mammals, 4.5% are birds, 45.8% are reptiles, 55.8% are amphibians and 33% are plants. These plants are endemic, being found nowhere else in the world.

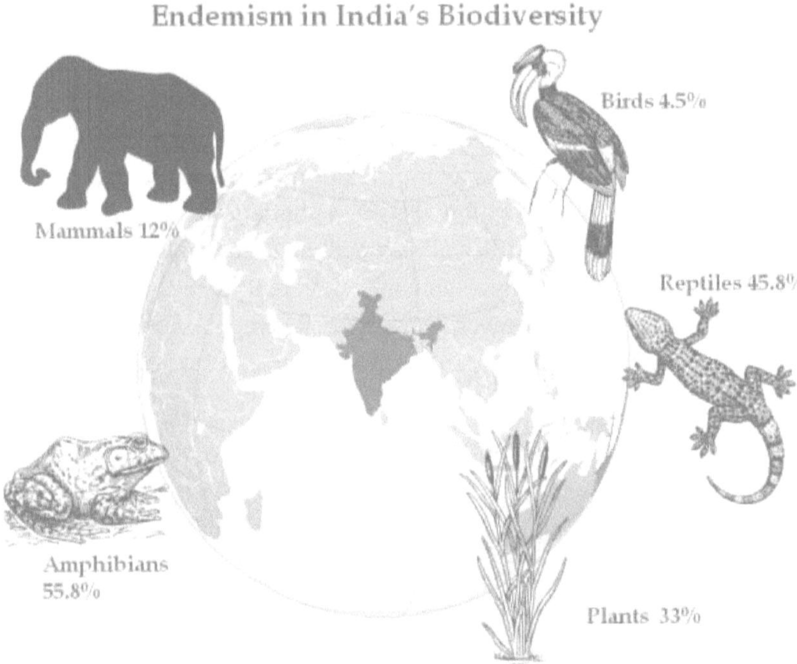

Figure 5.2: ***Represents a species-wise endemic content of the biodiversity in India.***

It is further estimated that about 400,000 more species may exist in India which need to be recorded and described. The baseline data on existing species and their macro-and micro-habitats, is also inadequate.

This biodiversity has arisen over the last 3.5 billion years of evolutionary history and its sustainable use has always been a part of the Indian culture. India is home to nearly one-fifth of the world's human population and is rapidly seeing a change in its economy from a predominantly agrarian society into a diversified one, resulting in mounting pressures on land use.

A consequence of this has been the loss and fragmentation of natural habitats, which has been identified as the primary threat to biodiversity.

India also has three of 34 "global biodiversity hotspots" - unique, biologically rich areas which are facing severe conservation threats. The rapid rate of hotspot degradation makes it imperative that conservation science be pursued immediately and vigorously in these habitats, to devise effective measures which curtail the rapidly diminishing biodiversity, and to protect its unique biota.

The value of this biodiversity for sustaining and nourishing human communities is immense. To take an example, the ecosystem services from the forested watersheds of two great mountain chains, the Himalayas and the Western Ghats, indirectly support several million people in India.

Open and free access to biodiversity information is essential to promote conservation, management and sustainable use of biodiversity and has immense potential to increase the current and future value of the country's biodiversity for a sustainable society.

5.3 Threatened species in India

In India, 44 plant species are critically endangered, 113 are endangered and 87 are vulnerable. Among animals, 18 are critically endangered, 54 are endangered and 143 are vulnerable.

5.3.1 IUCN Red List:

The IUCN Red List is a catalogue of taxa that are facing the risk of extinction. The uses of the Red list are:

1. Developing awareness about the importance of threatened biodiversity.
2. Identification and documentation of endangered species.
3. Providing a global index of the decline of biodiversity.
4. Defining conservation priorities at the local level and guiding conservation action.

The World Conservation Union has recognized eight red list categories of species: Extinct, Extinct in the wild, Critically endangered, Endangered, Vulnerable, Lower risk, Data deficient, and not evaluated.

5.3.1.1 Red list category Definition

5.3.1.1.1 Extinct: A species is extinct when there is no reasonable doubt that the last individual has died. Example: Dodo, passenger pigeon

5.3.1.1.2 Extinct in the wild: A species is extinct in the wild when exhaustive surveys, in known and/or expected habitats have failed to record an individual. Examples: Alagoas and curassow

5.3.1.1.3 Critically endangered: A species is critically endangered when it is facing an extremely high risk of extinction in the wild in the immediate future. Example: Gharial

5.3.1.1.4 Endangered: A species, whose numbers are so small that the species is at risk of extinction. Examples: Giant panda, Snow leopard, Tiger, and Indian rhinoceros.

5.3.1.1.5 Vulnerable: A species is vulnerable when it is not critically endangered or endangered but is facing a high risk of extinction in the wild in the medium-term future. Examples: Cheetah, Lion, and Polar bear.

5.3.1.1.6 Lower risk: A species is at lower risk when it has been evaluated and doesn't satisfy the criteria for critically endangered, endangered or vulnerable. Examples: Blue-billed duck and Solitary eagle.

5.3.1.1.7 Data deficient: A species is data deficient when there is inadequate information to make a direct or an indirect assessment of its risk of extinction.

5.3.1.1.8 Not evaluated: A species is not evaluated when it has not yet become assessed against the above criteria.

	Critically endangered	**Endangered**
Animals	 Pygmy Hog	 Asian Elephant
	 Namdapha flying squirrel	 Lion-tailed Macaque

	 Malabar large spotted civet	 Red Panda
Plants	 Berberis nilghiriensis	 Capparis pachyphylla
	 Dipterocarpus grandiflorus	 Pterocarpus santalinus
	 Ilex khasiana	 Madhuca diplostemon

Figure 5.3: ***List of few critically endangered and endangered species of India.***

5.4 National Policy for Biodiversity Protection

The Biological Diversity Act, 2002 (National Biodiveristy Authority, nbaindia.org) is an act to provide for conservat ion of biological diversity, sustainable use of its components and fair and equitable sharing of the benefits arising out of the use of biological resources, knowledge and for matters connected therewith or incidental thereto. It is administered by the National Biodiversity Authority (NBA), a statutory autonomous body headquartered in Chennai, under the Ministry of Environment and Forests, Government of India. The NBA was established in 2003 to implement the provisions of the Act. State Biodiversity Boards have been created in 29 states along with 31,574 Biological Management Committees (for each local body) across India.

The functions of the NBA include:

- Regulation of acts prohibited under the Act;
- Advise the government on conservation of biodiversity;
- Advise the government on selection of biological heritage sites; and
- Take appropriate steps to oppose grant of intellectual property rights in foreign countries, arising from the use of biological resources or associated traditional knowledge.

5.5 Activities

1. Observe the organisms in your school or garden, write their characteristics e.g. Spotted Frog, Green Parrots, Black Ants with their scientific names, food habits and habitats.
2. Write the folklores about 5 different mythological animals that are popular in Indian culture.

Spotted Frog.

Mythological animals mentioned in sacred texts.

3. Find the biodiversity of your school.

   Identify an area and observe which birds, animals, plants or insects are inhabiting that area.

   Make a map and label it with the names of the organisms that are observed. Be careful, not to harm the living beings that you are observing!

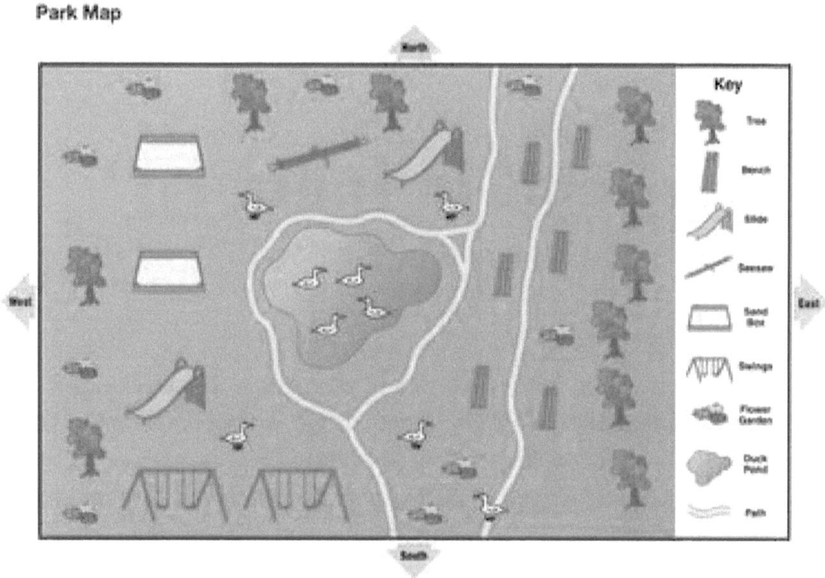

Mapping of the school area or garden for bio-diversity.

4. Research the National Biodiversity Act, and write a report on how you would use it to preserve and enhance the biodiversity in the western and eastern ghats, and the Himalayan region.
5. Research the biodiversity in the western ghats, which is considered a World biological diversity hotspot. Write a 5-page report, with appropriate graphics.
6. How many National Forests are there in India? Are any of them close to your city? Write a short report on the importance of the preservation of biological diversity in India.
7. Observe the biological diversity in your city park for one week, and write a report of your findings. Be sure to include all the flora and fauna, including the varieties of butterflies.

References

1. Source: https://www.nationalgeographic.org
2. Source: https://www.burkemuseum.org
   Source: https://www.speakingtree.in
   National Biodiversity Act, www.nbaindia.org

Chapter 6

Ecosystems in India

Introduction

***An ecosystem** is the living community of plants and animals in any area together with the non-living components of the environment such as soil, air and water, which constitute the ecosystem.*

An ecosystem includes all of the living things in a given area, interacting with each other and also with the non-living environment. Ecosystems are the foundations of the biosphere and determine the health of the earth system. Consider a puddle at the back of your home: in it you may find different types of living things from microorganism to plants to insects. These may depend on non-living things such as temperature, pressure and nutrients in water for their life. These interactions between living beings and their environment form the foundation for energy flow and recycle of carbon and nitrogen.

India is home for many endemic species and is ranked sixth among the 12 mega-diversity centers in the world. The 28 distinct biogeographic provinces, and the variety of life zones and floral groups results in diverse vegetation and ecosystems. In terms of plant diversity, India ranks tenth in the world and fourth in Asia. India has over 45,500 plant species, which is around 11% of the world's known floral diversity.

Ecosystems can be divided into terrestrial and aquatic ecosystems. Terrestrial ecosystems include Himalayan mountain, northern plains, peninsular plateau, deserts, coastal plants, island, tropical wet evergreen forests, tropical deciduous forests and scrubs, montane forests and alpine forests. Aquatic ecosystems include corals, estuaries, lakes, marine, mangroves, wetlands and rivers. The details on types of ecosystems are given later in this Chapter.

6.1 Understanding Ecosystems

As we discussed before, ecosystems involve interaction between living and non-living components of the environment. An ecosystem consists of a biological community that occurs in some locale, and the physical and chemical factors that make up its non-living or abiotic environment. There are many examples of an ecosystem, pond, forests, river, grassland, deserts etc. Study of ecosystems involves the study of certain processes linking living or biotic components to non-living or abiotic components. Two main processes that are mainly studied by scientists are energy transformation and the biogeochemical cycle. Study of ecosystems is known as Ecology. One can study ecology at the level of an individual, population, community and the ecosystems.

6.1.1 Components of an Ecosystem

The parts of an ecosystem are listed below, under the headings "abiotic" and "biotic".

Table 6.1: *The biotic and abiotic components in a typical ecosystem.*

ABIOTIC COMPONENTS	BIOTIC COMPONENTS
Sunlight	Primary producers
Temperature	Herbivores
Precipitation	Carnivores
Water or moisture	Omnivores
Soil or water chemistry (e.g., P, NO_3, NH_4)	Detritivores

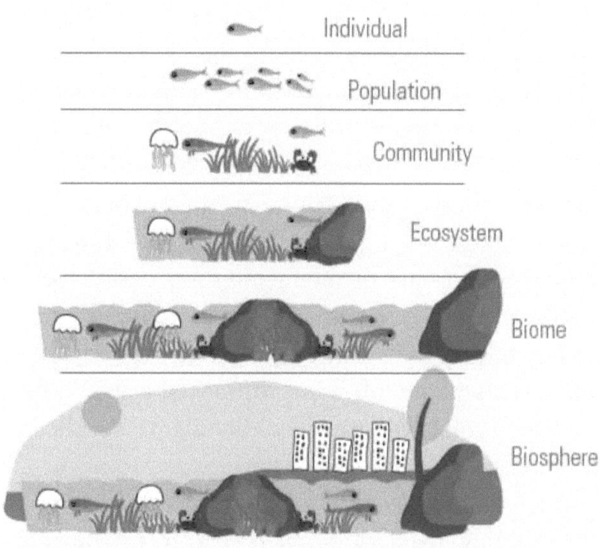

Figure 6.1: ***Levels of Organization in a biosphere.***

An individual is any living thing or an organism, population is a group of individuals of a given species living together at a specific geographical area at a given time. A community includes all the population living at the specific area at the given time, ecosystem is more than a community of a living organism interacting with the environment. A biome in simple terms, is a set of ecosystems sharing similar characteristics with their abiotic factors adapted to their environments. A biosphere is the sum of all the ecosystems established on planet Earth, it is the living (and decaying) component of the earth system.

In an ecosystem ecology, all the above things are put together and we try to understand how the ecosystems works. This means that, rather than worrying mainly about particular species, we try to focus on major functional aspects of the system. These functional aspects include such things as the amount of energy that is produced by photosynthesis, how energy or materials flow along the many steps in a food chain, or what controls the rate of decomposition of materials or the rate at which nutrients (required for the production of new organic matter) are recycled in the system.

6.2 Influence of Humans on Ecosystems

The growing human population has exerted a significant impact on all ecosystem types in all regions of India. High human and livestock demand for food, fodder and firewood has resulted in overexploitation of terrestrial as well as aquatic ecosystems. Wind and water erosion, water logging and problems of salinity and alkalinity have resulted in land degradation of at least one third of the geographical area. Human activities have also resulted in high air and water pollution. India now has 321 globally threatened floral and 614 faunal species as per the IUCN (2012) Red List.

There has been a conspicuous change in land use patterns, habitat loss and forest fragmentation. Large tracts of natural forests have been converted into cultivated land or monoculture of cash and timber crops such as tea, coffee, rubber, teak and eucalyptus. River systems have been greatly affected by building dams and reservoirs. The wetland ecosystems are threatened by discharge of waste effluents, surface runoff, weed infestation and uncontrolled siltation. Overfishing has resulted in a decline of fishery resources of fresh and marine water bodies. Changes in temperature, precipitation, evapo-transpiration and run-off have been responsible for altering the hydrological regime of the inland natural wetlands, particularly in arid and semi-arid regions (Patel et al., 2009).

On account of anthropogenic factors (reclamation of land, discharge of wastes, etc.) and natural factors like global warming, mangroves are presently one of the most threatened ecosystems (MoEF, 2012). Due to diversion of freshwater in the upstream area, the periodicity and quantity of freshwater reaching the mangrove environment has been reduced leading to a dramatic change in the floral diversity of the mangrove wetlands (Selvam, 2003). India is taking significant steps to conserve the biodiversity of the country.

6.3 Ecosystem Processes

Figure 6.2 shows how an ecosystem functions, energy flow and the material cycle are linked together but are not the same. Energy enters the biological systems in the form of light energy or photons, they are converted to chemical energy by the biological processes such as photosynthesis and respiration, energy released during this process is lost to the environment in the form of heat, which cannot be recycled. Elements such as phosphorus, carbon, nitrogen enter into the living systems from the surrounding atmosphere, soil and water. They are transformed biochemically within the organisms' body but later due to decomposition or mineralization, returned to an inorganic state (phosphorous, nitrogen, carbon). The elements are cycled endlessly between their biotic and abiotic states within ecosystems. Those elements whose supply tends to limit biological activity are called nutrients.

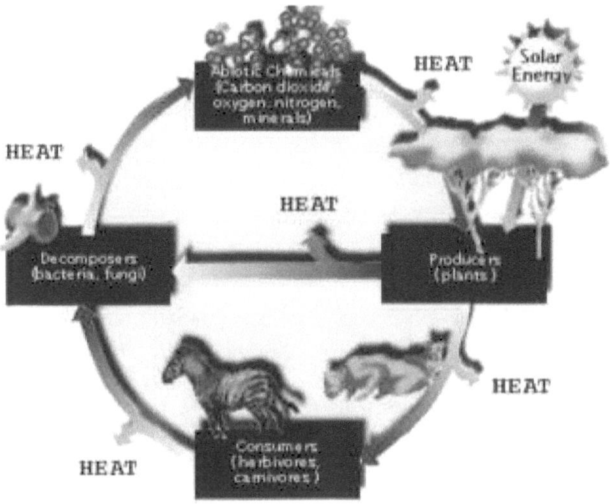

Figure 6.2: ***Energy flow and material cycle.***

6.4 India and its Ecosystems

India occupies approximately 2.5% of the world's surface and is home to 7% of the world's biodiversity. India's species and genetic diversity is spread across its ecosystems. India's climate and its habitat that ranges from plains to wetlands to mountains in the north, has contributed to its biodiversity across diverse ecosystems. The major types of ecosystems observed in the Indian subcontinent are as follows:

- Terrestrial
- Aquatic

6.4.1. Terrestrial Ecosystems :

'Terrestrial' as the name suggests, these ecosystems typically occur on land whose inhabitants encompass a wide range of flora and fauna .

India is home to the following forms of terrestrial ecosystems:

i. Forests
ii. Grasslands
iii. Dessert
iv. Mountains

6.4.1.1 Forests: India has a wide variety of forests owing to its geography viz. Tropical, Sub-tropical, Temperate and Alpine.

6.4.1.1.1 Tropical Forests: These types of forests occupy most of India's landmass, except regions of high altitudes. Also known as India's rainforests, **tropical evergreen forests** can be seen across the Western Ghats, plateaus, Andaman and Nicobar Islands, upper Assam, Odisha coast and lower eastern slopes of the Himalayas.

Figure 6.3: ***Tropical Evergreen Forest in India.***

6.4.1.1.2 Tropical deciduous forests: are found in major parts of Kerala, Madhya Pradesh and Maharashtra. Dry deciduous forests are found in regions of West Bengal, Gujarat and Rajasthan.

Figure 6.4: ***Tropical Deciduous Forest.***

6.4.1.1.3 Tropical Dry Forests: are mainly found in the Northern Hills and some parts of South India. The trees in these regions shed their leaves in the advent of winter and become lush green again after winter is over.

Figure 6.5: ***Tropical Dry Forest.***

6.4.1.1.4 Sub-Tropical forests, especially hill forests, reign the Nilgiris and parts of the Himalayas mainly at an altitude of approximately 2000 meters. Sub-tropical pine forests are found mainly at higher altitudes of North Eastern hilly terrains of Assam, Meghalaya, Nagaland & Mizoram.

Figure 6.6: ***Subtropical Pine forest.***

6.4.1.1.5 Montane Temperate Forests are found abundantly in the Northern middle Himalayas viz. Himachal Pradesh, Uttarakhand and Sikkim.

Figure 6.7: ***Temperate Forests of the Himalayas.***

6.4.1.1.6 Alpine Forests can be seen below the snowline above the altitude of 3000 meters. Vegetation in these forests is rare and mainly consists of meadows and grasslands.

Figure 6.8: ***Alpine Forest of Himalayas.***

6.4.1.2 Grasslands: These ecosystems constitute 24% of India's geography and are extremely dynamic in nature in the sense that they encompass all natural pastures, woodlands, and steppe formations within the Himalayan ranges. They comprise of grasses and grass-like vegetation. Grasslands can be seen across the country in the Alpine regions of the Himalayas, the Gangetic and Brahmaputra plains, hill top areas of Western Ghats, dry regions of Andhra Pradesh ,western and peninsular regions, floating grasslands of Manipur, and North Eastern Hilly regions. These grazing grounds with patches that contain ponds and rivers are easily accessed by the rural folk. These grasslands are a source of food, water, fodder, firewood and hence the source of livelihood to rural folk living in these regions. These geographical structures help maintain groundwater, contributing to ecological balance.

Figure 6.9: ***Grassland s of India.***

6.4.1.3 Deserts: Deserts are geographic land formations that are a result of a very slow process of land degradation. Scarce vegetation and very high evaporation rates are a characteristic of deserts.

India is home to the 7th largest desert in the world, Thar desert. Sand dunes are a peculiar characteristic of this desert while other parts of the desert comprise of grasslands with scanty vegetation, and dry salt lake beds.

Rann of Kutch in Gujarat is known as the white desert for the layers of white salt it is covered in. Temperatures can soar to 50 °C in summers in this region and dip below freezing during winters.

Cold desert of Spiti Valley in the northernmost part of the country bears witness to extreme temperature (freezing) and very less rainfall. Inhabitants in this region are sparse and is home to rare wildlife.

Figure 6.10: ***Thar Desert.***

Figure 6.11: Rann of Kutch.

6.4.1.4 Mountains: India has 2 main mountain ranges, the Himalaya and the Western Ghats. Himalayas flank the northern and north eastern regions of the country, covering Jammu & Kashmir, Himachal Pradesh, parts of

Uttar Pradesh, Sikkim, Arunachal Pradesh, Nagaland, Manipur, Mizoram, Tripura, Meghalaya, and a small part of Assam.

6.4.1.4.1 Himalayas are a mix of glaciers, deep valleys and fertile terrain. Nandadevi National Park is designated as a World Heritage site by UNESCO. It harbours 5 climatic regions viz.Tropical,Sub-Tropical, Temperate, Alpine and Arctic. It is one of 25 hotspots in the world and also the origin of 3 main rivers, Ganga, Brahmaputra and Indus.

Figure 6.12: ***Himalayas.***

6.4.1.4.2 The Western Ghats is one of the 8 hottest hotspots in the world and stretches from the mouth of river Tati in the North (Gujarat) across the entire western coast of India to the south of India only to be separated by a short gap known as the gap of Palghat.

The Ghats run perpendicular to the western coastline as a result of which moisture-laden south-west monsoon from the Arabian Sea, are forced to cross this range to get to the other side. This climb leads to decrease in pressure in the atmosphere which in turn leads to cooling. Condensation of the moisture occurs which leads to an annual rainfall of 2,000 mm to 7,800 mm .

6.4.2 Marine Ecosystems

Marine ecosystems are abundant in the Arabian Sea and Bay of Bengal. The various islands in the Arabian Sea and Bay of Bengal have their unique ecosystems. The Andaman and Nicobar Islands of Bay of Bengal and the Lakshadweep islands in the Arabian Sea have flora and fauna that attract a lot of tourists from around the world. The Andaman and Nicobar Islands are much larger than Lakshadweep, and cover an area of 8,154 sq. Km with a population of 380,000. There are over 300 islands with palm-lined, white sand beaches. Mangroves and tropical rain forests cover the islands. Coral reefs support a variety of marine life and sharks and other large mammals are found in the water around the islands.

Lakshadweep is a group of 36 islands covering an area of 32.7 sq. Km. The population is about 65,000 people. 10 of the islands are inhabited. Sun-kissed beaches and lush greenery is found in the islands. There are 12 atolls, 3 major reefs and 5 submerged banks in Lakshadweep.

6.5 Activities and Exercises

1. Select one of the ecosystems described in this chapter and do your own research to get additional information. Write a 5-page report, with illustrations, and describe how this ecosystem is being affected by human activities.
2. Study the flora and fauna of the Andaman and Nicobar islands and prepare a 5-page report with illustrations. When and why was the Andaman Island used as a penal colony by the British Government?
3. Research the Lakshadweep islands and describe the types of marine life in the waters surrounding the islands. Has tourism impacted the quality of the environment? If so, write a report about the environmental impacts.
4. Why is the Western Ghats considered one of the 8 biodiversity hotspots in the world? Describe the flora and fauna in the western ghats, and how is the government protecting the biodiversity of the region?
5. Describe the origin of the deserts of Rajasthan and the Rann of Kutch. How do the people live in such an arid land and how have they constructed rain water harvesting systems to capture the minimal rainfall in the region? Write a 5-page report about your findings.
6. What are the ecosystems that support the people and flora/fauna in the Himalayan mountainous region? How is Climate Change impacting the glaciers in the Himalayas? Write a 5-page report, detailing what can be done to minimize the melting of the glaciers in the region.
7. What is your favourite ecosystem in India, based on the ecosystems described in this chapter? Explain your answer with a 5-page report.

References

1. http://officersiasacademy.blogspot.com/2016/03/ecosystems-of-india.html]
2. https://www.researchgate.net/publication/319876388_Diversity_of_Ecosystem_Types_in_India_A_Review]
3. http://indo-germanbiodiversity.com/children-segment/ecosystems-in-india.php]
4. http://eschooltoday.com/science/ecosystems/levels-of-organisation-in-an-ecosystem.html
5. https://globalchange.umich.edu/globalchange1/current/lectures/kling/ecosystem/ecosystem.html
6. https://www.researchgate.net/publication/319876388_Diversity_of_Ecosystem_Types_in_India_A_Review
7. https://globalchange.umich.edu/globalchange1/current/lectures/kling/ecosystem/ecosystem.html
8. http://www.biologydiscussion.com/india/biodiversity-india/biodiversity-in-india/70823
9. https://sciencing.com/types-terrestrial-ecosystems-5516822.html
10. http://frienvis.nic.in/KidsCentre/Types-of-Indian-Forest_1811.aspx
11. http://www.yourarticlelibrary.com/environment/forest/forests-types-top-4-types-of-forests-found-in-india/74987
12. http://wiienvis.nic.in/WriteReadData/Publication/19_Grassland%20Habitat_2016.pdf
13. https://www.downtoearth.org.in/news/agriculture/india-lost-31-of-grasslands-in-a-decade-66643
14. https://www.eolss.net/Sample-Chapters/C20/E6-142-DE-04.pdf
15. https://sciencing.com/list-deserts-india-7477131.html
16. https://www.cbd.int/doc/world/in/in-nr-me-en.pdf
17. https://www.thehindubusinessline.com/news/science/western-ghats-biodiversity-is-a-significant-source-of-moisture-for-monsoon/article23772839.ece
18. Patel et al., 2009

Chapter 7

Natural Hazards and Disaster Management

Introduction

Natural hazards that frequent India are floods, droughts, cyclones, landslides and avalanches in the mountainous terrains of the Himalayan foothills. Of these, floods and droughts result in death and property damage on a large scale almost every year, and the government of India is trying to predict these accurately and reduce the damage caused by flooding. In addition, efficient management of the many large river basins can minimize the impacts of flooding and droughts. Apart from the monsoons, rapid snowmelt in the Himalayan region produces significant flooding in northern India. When the monsoons are delayed or the amount of rain produced by the monsoon storms is reduced, many areas of the country experience drought. Rain-fed agriculture is practiced in many parts of the country and the lack of rain produces drought in many parts of India, especially in the desert northwest of the country.

This chapter will describe the impacts of floods and droughts, earthquakes, cyclones, landslides, and snowstorms/avalanches in India. The worsening of some of these disasters due to climate change will also be discussed. Disaster management to alleviate the impact of these natural hazards will be described, and how this can be improved with accurate predictions of these events will also be discussed. Finally, several Activities and Exercises are given in the end to challenge the students and review the student's learning of the subject matter.

7.1 Floods and Droughts

Severe weather and infrastructure that is inadequate for expected weather conditions cause severe flooding in urban areas. This causes severe economic losses due to property damage and loss of agricultural crops, and death of humans and animals. In recent years, due to severe weather warning from satellites, the economic losses and death have been minimized. However, weather forecasting is still in an inexact science, and the loss of property and lives is an annual occurrence. Various types of insurance policies help alleviate the economic losses, but disruption of lives is an ongoing occurrence. Maintaining the floodplains of rivers free from valuable property and allowing a flood surge zone in coastal areas is a flood control technique followed in the Netherlands and Denmark, and they have minimized property damage and loss of life due to floods that happen in their countries. Enforcement of strict land-use zoning by local, State and Central Government authorities is required to ensure that valuable properties are not built in flood prone zones and in coastal areas where the flood surges frequently occur due to severe weather conditions.

India has several large river basins and these are fed by snowmelt from the Himalayas in the north and by the monsoon rains in the rest of the country.

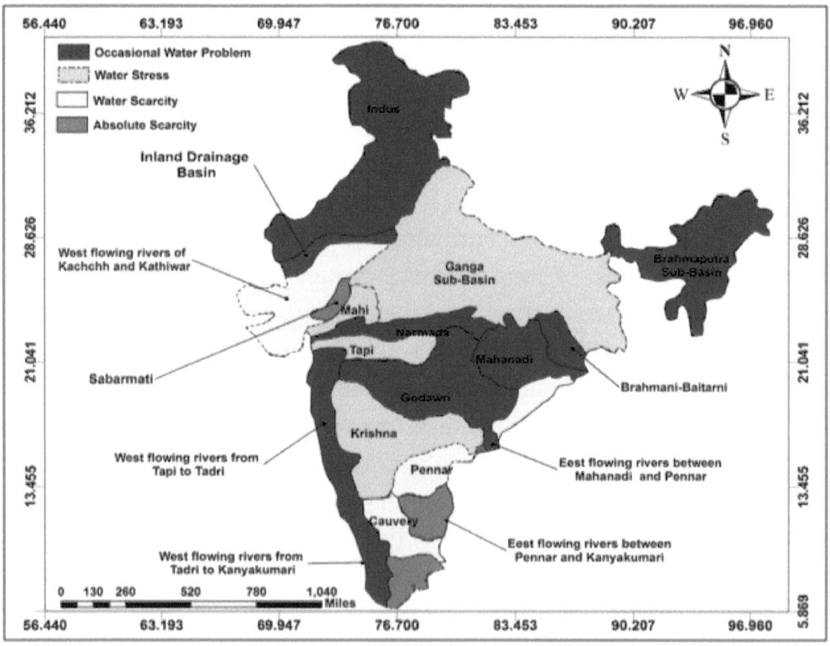

Figure 7.1: ***India's river basins.***

The Brahmaputra River in the northeast is a river that floods annually with heavy rains that add to the water from the snow melting in the

Himalayas. Flood plain assessment and proper planning could alleviate the frequency and damage resulting from this flooding. The Ganges River floods frequently and since the banks are heavily developed along the many cities where the River flows, the property damage and loss of lives is significant. Monsoon rains provide the water for most of the river basins in India, and if the monsoons are severe in a year, there is a lot of flooding, and in years when the monsoons are light, there is drought in many parts of the country.

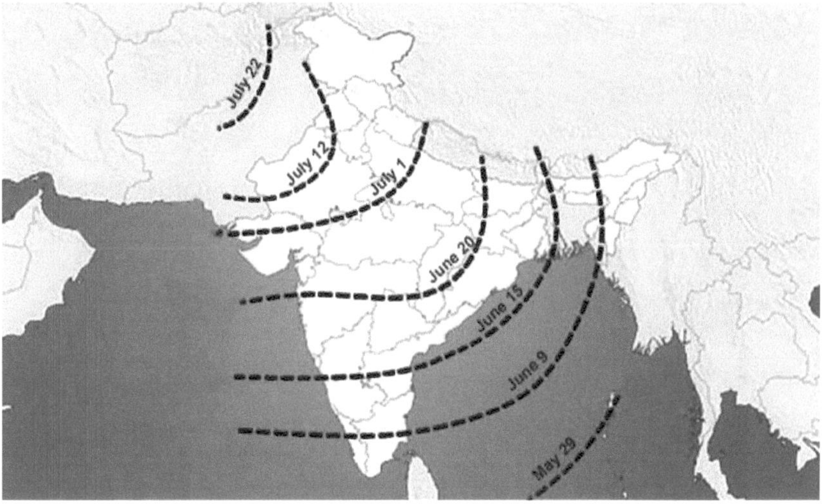

Figure 7.2: ***Monsoon patterns in India.***

Although weather patterns and snow melt are hard to predict, one can take the following steps to minimize the impacts of floods and droughts:

7.1.1 Urban areas:

1. Develop emergency notification/flood warning procedures for populations affected by flooding;
2. Promote green infrastructure in all areas of the city, starting with government buildings so that rainwater capture is maximized. Techniques available include bioswales along streets, cisterns (above and below ground) to capture the rainwater from buildings, creation of open space, cleaning the debris from storm drains, recharge pits, and roof gardens.

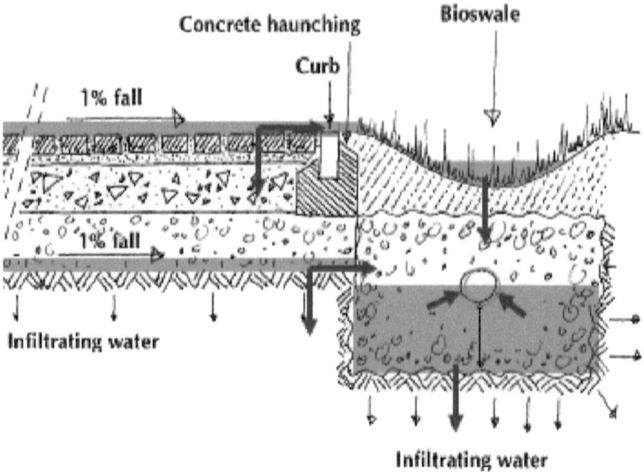

Figure 7.3: ***Bioswales.***

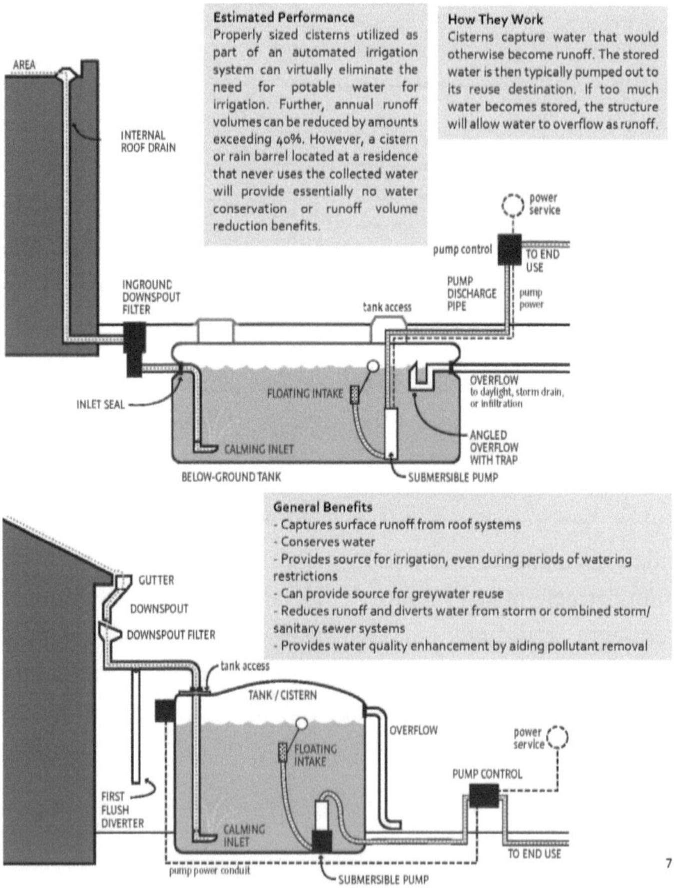

Figure 7.4: ***Cisterns.***

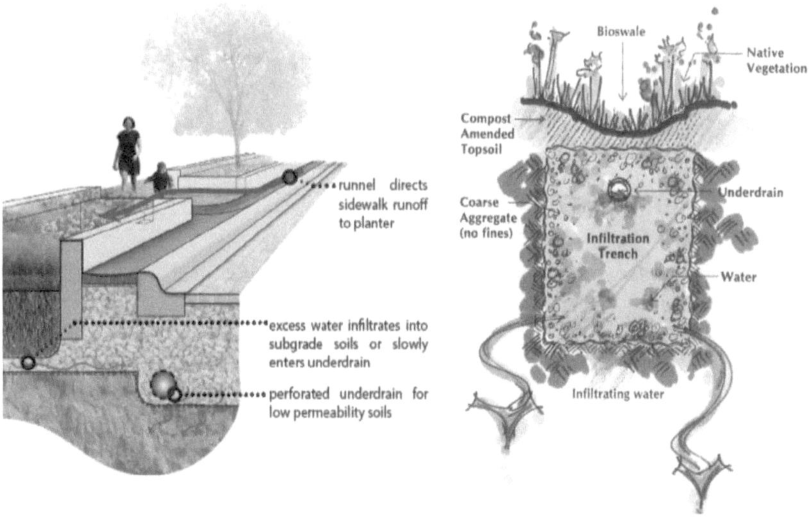

Figure 7.5: ***Recharge pits.***

Figure 7.6: ***Rain gardens.***

3. Promote water conservation in all parts of the city;
4. Repair leaks in stormwater, wastewater and drinking water pipes. These will minimize waste of water and serve well in drought conditions; and
5. Keep river bank areas free from development so that property damage is minimized during river floods.

7.1.2 Rural Areas:

1. Keep floodplains of rivers free from residential development so that flood waters do not damage property and lives;
2. Build reservoirs for monsoon water storage so that probability of droughts is minimized;
3. Build recharge pits in key locations so that rain water is captured and flooding is minimized; and
4. Prevent excessive pumping of groundwater so that it is available during periods when the monsoon are not as much as predicted and drought conditions prevail.

In addition to these measures, timely evacuation procedures for people and shifting movable property to safer grounds is necessary to minimize loss of life and property.

As global temperatures are rising due to climate change, as much as 5% of the world's population will be flooded every single tear by 2100, if we do not halt our greenhouse gas emissions (Wallace-Wells, 2019). In 2016, climate scientist James Hansen had suggested sea level could raise several meters over fifty years, if ice melt doubled every decade. However, actual measurements of ice sheets in Antarctica and the Arctic indicate ice melt faster than this. The combination of coastal flooding due to sea level rise and inland flooding due to rising river levels due to heavy rains will devastate property and lives in many parts of the world. Wallace-Wells (2019) states that India will be hit severely with climate change, shouldering nearly twice as much as the burden of the next nation.

7.2 Earthquakes

Earthquakes are relatively rare in India, except in the Himalayan region and Northeastern parts of India where they occur with some frequency.

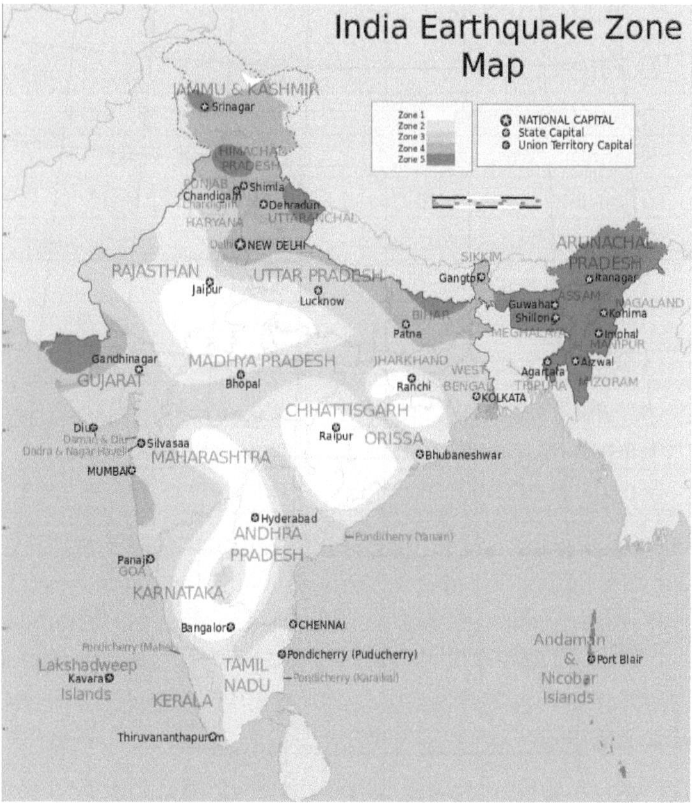

Figure 7.7: ***Earthquake zones in India.***

Earthquakes occur when there is a sudden movement along faults (also called tectonic plates) below the earth's surface, either on land or in the ocean. There are many small earthquakes happening in many parts of the world where there are active faults; however, they cause disasters only when they occur in populated areas and when the intensity of the earthquake is above 5 or 6 on the Richter Scale (used to measure the intensity of earthquakes). Large magnitude earthquakes below the ocean's surface disturb the water surface and create large damaging waves called Tsunamis (Bhargava et al, 2016). Such a tsunami affected Indonesia, Thailand and the southeastern coast of India in December 2004. This tsunami destroyed property worth several billions of dollars and killed thousands of people in several countries. Japan has suffered many tsunamis, the last one in Fukushima Japan severely damaged a nuclear power plant and caused the spread of radiation for many miles surrounding the plant.

The ten worst earthquakes in India are:

Table 7.1: *Major earthquakes that rocked India*

No.	Location	Date
1.	Indian Ocean tsunami	December 26, 2004
2.	Kashmir earthquake	October 8, 2005
3.	Bihar and Nepal earthquake	January 15, 1934
4.	Gujarat (Bhuj) earthquake	January 26, 2001
5.	Kangra (Himachal Pradesh)	April 4, 1905
6.	Latur (Maharashtra)	September 30, 1993
7.	Assam	August 15, 1950
8.	Assam	June 12, 1897
9.	Garhwal (Uttarakhand)	October 20, 1991
10.	Koynanagar (Maharashtra)	December 11, 1967

These earthquakes ranged in magnitude from 6.4 to 9.3 on the Richter Scale, with the Indian Ocean earthquake/tsunami being the worst. The loss of life and property damage was considerable in these earthquakes.

The Indian government is learning from these earthquakes and has taken several precautions and preparedness drills to minimize loss of lives and property damage. These include:

- Early warning programme based on field monitoring devices and rugged and reliable communication system;

- Earthquake-resistant construction using materials and methods that will allow the concrete building or pavement to move but not collapse;
- Planning of utilities in a manner that they will not be destroyed during the earthquake, and any fires resulting from them can be controlled;
- Emergency preparedness procedures that will respond to the collapse of structures and other emergencies within the shortest time possible; and
- Trained personnel in hospitals and government who can treat the injured and manage disruptions to the normal life in a community.

Institutions in India are studying the geologic features that result in earthquakes and are constantly trying to improve the monitoring of movement along faults in the Hmalayan region and in the coastal mountain ranges. The Geological Survey of India and the Indian Institute of Earthquake Engineering in Roorkee, India, are working with Central and State governments to improve the prediction of earthquakes and building earthquake-resistant structures. The sophistication of earthquake monitoring has improved significantly worldwide, and there is significant international cooperation in this field. Governments from around the world respond to earthquake catastrophes to minimize loss of life and help in rebuilding communities.

7.3 Cyclones

The tropical cyclones are rapidly rotating storm systems characterized by a low-pressure center, a closed low-level atmospheric circulation, strong winds, and a spiral arrangement of thunderstorms that produce heavy rain. These form exclusively over tropical areas of the Indian Ocean. These are called hurricanes in the North Atlantic Ocean, South Pacific Ocean, or North-east Pacific Ocean. In the northwest Pacific Ocean west of the International Date Line, they are called typhoons. Cyclone refers to their winds moving in a circle, whirling around the central clear eye, with their winds blowing counter-clockwise in the northern hemisphere and clockwise in the southern hemisphere (Wikipedia, Tropical Cyclone, 2018). Tropical cyclones typically form over large bodies of relatively warm water. They derive their energy through evaporation of water from the ocean surface, which ultimately re-condenses into clouds and rain when the moisture rises and cools to saturation. As global warming increases, the temperature of the water raises and increases the intensity of the cyclones.

Tropical cyclones are generally between 100 and 2000 km in diameter (Wikipedia, Tropical Cyclone, 2018). The strong rotating winds are a result of the conservation of angular momentum imparted by the Earth's rotation as air flows inwards toward the axis of rotation. Wind speeds are more than 74 miles per hour (119 km/hr), and their intensity is rated on the Saffir-Simpson scale (Bhargava et al, 2016). They rarely form within 5 degrees of the equator. Coastal regions are particularly vulnerable to the impact of cyclones, compared to the inland regions. The energy of the cyclone is from the warm ocean waters, and so as the cyclone moves inward onto land, they weaken quite rapidly. Coastal damage is caused by strong winds and rain, high waves, storm surges (due to wind and severe pressure changes), and the potential for spawning tornadoes. Tropical cyclones draw in air from a large area, and concentrate the precipitation of the water content in that air into a much smaller area, causing extremely heavy rain and river flooding up to 40 km from the coastline. Though the potential damage from cyclones is severe, they can relieve drought in areas around the coastline. Cyclones have a major impact on regional and global climate.

There are six Regional Specialized Meteorological Centers (RSMCs) worldwide to monitor and predict cyclones. These organizations are designated by the World Meteorological Organization and are responsible for tracking and issuing bulletins, and advisories about tropical cyclones in their designated areas of responsibility. In addition, there are six Tropical Cyclone Warning Centers (TCWCs) that provide information to smaller regions. The RSMCs and TCWCs are not the only organizations that provide information about tropical cyclones to the public. The Joint Typhoon Warning Center and the Philippine Atmospheric, Geophysical and Astronomical Services issue advisories. The table below provides the Tropical Cyclone Basins and Official Warning Centers (Wikipedia, 2018).

Table 7.2: *Tropical Cyclone Basins and Official Warning Centers*

Basin	Warning Center	Area of Responsibility
North Atlantic	US National Hurricane Center	Equator northward, African Coast 140 W
Eastern Pacific	US Central Pacific Center	Equator northward, 140 Deg. W-180
Western Pacific	Japan Meteorological Agency	Equator -60 deg., 18-100 E

Basin	Warning Center	Area of Responsibility
Northern Indian Ocean	India Meteorological Department	Equator northward, 100-45 E
South West Indian Oc.	Meteo-France Reunion	Equator 40 deg. S, African Coast 90 E
Australian Region . .	Australian Bur. Of Met Indonesian Agency for Met. Papua New Guinea NWS	10 deg. S-36S, 90-160 E Equator 10S, 90-141 E. Equator 10 deg. S, 141-160 E.
Southern Pacific	Fiji Met. Service Met. Service of New Zealand	Equator – 25 S, 160 E – 120 W 25 deg. S-40 S, 160 E – 120 W.

This shows that a lot of international cooperation is needed to predict and warn populations of impending tropical cyclones to minimize property damage and loss of lives.

Worldwide, tropical cyclone activity peaks in late summer, when the difference between temperature aloft and sea surface temperature is the greatest (Wikipedia, 2018). However, each basin has its own seasonal patterns. May is the least active, September is the most active, and November is the only month when all the tropical cyclone basins are active. Many factors affect the formation of cyclones, and include water temperatures in the first 160-foot depth of water, rapid cooling with height above the water surface, humidity, wind shear, and a preexisting system of tropical disturbances.

On October 9, 2013, a severe cyclone named Phaillin landed in Odisha, India, and around 12 million people were affected. This was the worst cyclone to hit India since the super cyclone that hit Odisha in 1999.

The Indian Meteorological Department is responsible for tracking and providing warning to concerned user agencies. Cyclone tracking is done through INSAT satellite and 10 cyclone detection radars. Warnings are issued to ports, fisheries, and aviation departments. The warning system provides for a cyclone alert of 48 hours, and a cyclone warning of 24 hours. There is a special disaster warning system for the dissemination of cyclone warning in local languages through INSAT to designated addresses in isolated places in coastal areas.

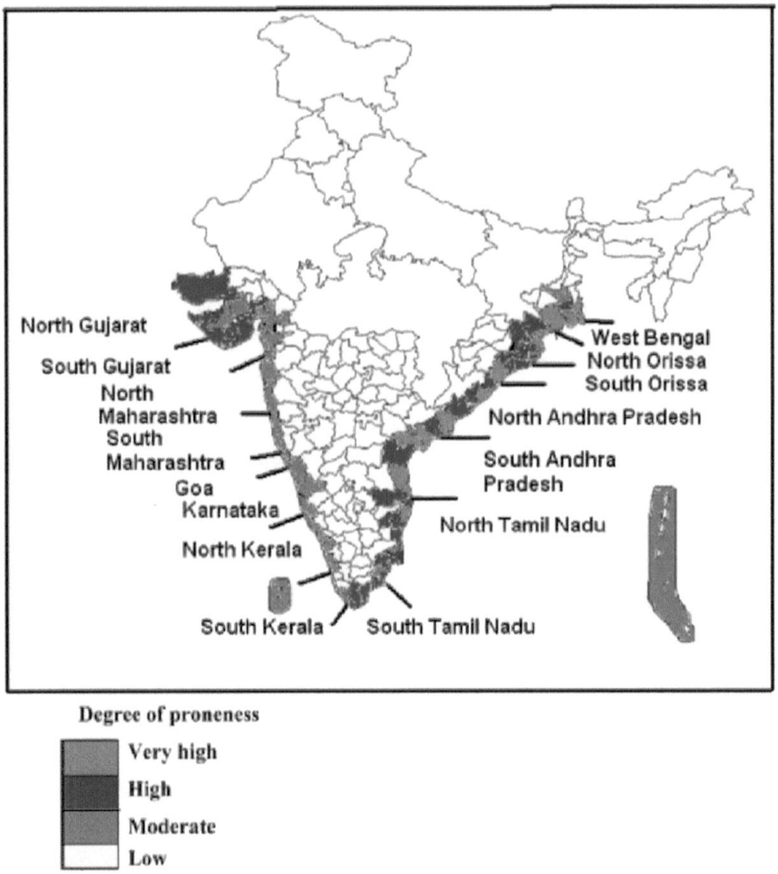

Figure 7.8: ***Cyclone zones of India.***

7.4 Landslides

Landslides are mass movements of soil or rock on unstable slopes and are natural phenomena. However, man-made actions such as building houses on steep slopes or undermining a soil/rock slope could create instability and cause a landslide. Landslides result in the loss of life and property damage in populated areas. Landslides are caused due to the following factors:

- Extremely weak zones in a soil or rock slope which causes sudden failure of the slope;
- Heavy rain which saturates a slope and causes the slope to fail;
- Active seismic or tectonic fault lines in a slope which fail due to an earthquake; and

- Man-made disturbances to a rock or soil slope which creates instability and produces slope failure.

To minimize landslides, the geology of an area should be studied and weak portions of the slope should be stabilized through rock or soil anchors. Slopes should be monitored, as is done in many road cuts and open pit mines, to predict developing unstable conditions and implement remedial measures. The following precautions should be undertaken to prevent loss of life and damage in landslide prone areas:

- Drainage channels along road and bridges to prevent saturation of the soils;
- Proper road alignment, avoiding weak planes, and old debris-filled areas;
- Growing trees on slopes which absorb the moisture and keep the slopes relatively dry;
- Design slopes which are stable after the geology is properly studied; and
- Implement remedial measures on steep slopes so that rock falls or mass soil failure is avoided.

In hilly and mountainous areas, slopes should be monitored on a frequent basis to determine if unstable conditions are being developed. An early warning system should be implemented so that conditions causing slope instability can be detected and remedial measures undertaken.

Figure 7.9: ***Landslide.***

A disaster management plan should be prepared and personnel trained to react quickly to landslides, and minimize loss of life and property damage. Local, State and Central government authorities should coordinate activities so that landslides are prevented and if they occur, to respond quickly and efficiently.

7.5 Snowstorms and Avalanches

Snowstorms cause huge economic disruptions and loss of life in Europe, North America and northern parts of India. In mountainous areas, heavy snowfall causes avalanches, which are defined as a sudden slide of a large mass of snow in a mountain slope.

Figure 7.10: ***Avalanche.***

With good meteorological data collection and analysis, prediction of snowstorms has improved and warning provided to communities. This helps manage the disastrous effect of these snowstorms and avalanches. In India, the areas prone to snowstorms and avalanches are Himachal Pradesh, Uttarakhand, and higher elevation in Kashmir, and the northeastern states.

A good disaster management plan minimizes the loss of life and property damage in a snowstorm. It involves early warning systems, and emergency preparedness so that trained personnel can be dispatched to help people affected by snowstorm and avalanches.

7.6 Wildfires

As temperatures rise around the world, the fire season is getting longer with devastating wildfires destroying lives and property in various parts of the world. The devastating fires in Australia made international news in early 2020, and the impact on wildlife was significant. Globally, since just 1979, the fire season has grown nearly 20%, and American wildfires now burn twice as much land as they did in 1970. By 2050, destruction from wildfires is expected to double again, and in some places within the United States the area burned could grow fivefold (Wallace-Wells, 2019). Globally, deforestation accounts for about 12% of carbon emissions, and forest fires produce as much as 25%. The combination of droughts and higher temperatures create the right environment for devastating wildfires in forests around the world. It is estimated that each year, between 260,000 and 600,000 people die from smoke from wildfires, and Canadian fires have been linked to spikes in hospitalizations as far away as the eastern coast of the United States (Wallace-Wells, 2019).

As urban centers get closer to forest lands, the property damage and lives lost due to wildfires increase significantly. Unless we reduce our demand for land and protect our forests, the impact of wildfires will keep increasing. In India, many poor communities live in or close to forested areas, and the impact of wildfires will be most directly felt by them. As we lose forested areas due to clear cutting of forests (happening in the Amazon in Brazil) or due to wildfires, the increase of greenhouse gases will threaten our futures with more drastic impacts due to climate extremes.

7.7 Disaster Management

Worldwide cooperation during natural disasters has minimized the loss of life and property damage resulting from the natural hazards mentioned above. Disaster management has two components, one, actions taken immediately after a disaster to rescue injured people and provide relief to the affected population, and two, long-term recovery of the community after the disaster. Usually, a lot of agencies help, including many non-governmental organizations, help in the rescue and immediate relief efforts but the long-term recovery is left to the local agencies and voluntary organizations. Since the tsunami of 2004, the Indian government set up the National Disaster Management Authority (NDMA) which has the following responsibilities:

- Prepare policies on disaster management;
- Approve a national plan for managing disasters;
- Approve plans prepared by the various agencies of the Central government, and the local/state governments;

- Formulate guidelines to be followed by state authorities in formulating the state plan;
- Coordinate the enforcement and implementation of the policy and plans;
- Recommend the funding for the purposes of rescue and recovery;
- Provide support to other countries affected by major disasters;
- Take measures for the prevention of disasters, mitigation, preparedness, and capacity building among local/state agencies; and
- Formulate broad policies and guidelines for the National Institute of Disaster Management.

With the active participation of NDMA, the loss of life and damage to property during disasters has been minimized.

Climate change is increasing the frequency and intensity of disasters. Globally, researchers have found an increase of 23 to 30% in Category 4 and 5 hurricanes for just one degree Celsius of global warming. Typhoons have intensified by between 12 and 15%, and the proportion of Category 4 and 5 storms has doubled, in some areas, it has tripled (Wallace-Wells, 2019). The paths of destructive tornadoes are getting longer and wider, impacting communities more frequently than earlier. For the world's poor, recovery from frequent wild storms is almost impossible. Their lives are disrupted forever and it is incumbent on governments to prepare for and provide for minimizing the devastation among poor communities.

7.8 Activities and Exercises

1. Research the Tsunami that happened in south India in 2004, and write a report of the damage caused to property, and loss of lives. How have the communities recovered from the Tsunami?
2. Analyze flooding in Brahmaputra River annually and prepare a graph of property damage due to flooding for the years 2010 to 2020. What measures have been taken by local, State and Central governments to reduce the impact of flooding and save lives?
3. How frequent are droughts in India and which states do they affect most? Graph the droughts in the years 2010 to 2010, and see if there is a pattern between them and the arrival of the monsoon rains in different parts of India.
4. Why do cyclones happen in the eastern coast of India? Analyze the cyclones that have affected Odisha from 1999 to date, and how has disaster management improved in the last 20 years.

5. Plot a graph of wildfires in India since 2000. Describe the impact of these fires on the lives of people and the damage to wildlife in affected areas.
6. How many avalanches happen in the northern states in the Himalayan foothills? Have they caused loss of lives and how have they impacted tourism in those states?
7. When was the National Disaster Management Authority (NDMA) set up in India? Analyze how NDMA works with the states to manage disasters in the states. Select one disaster and write a 5-page report on how the aid was delivered for emergency assistance and for long-time recovery efforts from the disaster.
8. Analyze the landslides in the states of Uttarakhand, Himachal Pradesh, and Ladakh in the year 2013 and write a report detailing the property damage, lives lost and how the recovery efforts since then.
9. Obtain a map of India showing the various earthquake-prone areas. Analyze the earthquakes in these areas for the period 3010 to 2020. Is India better prepared for earthquakes in 2020 compared to the Bhuj earthquake in Gujarat? What steps have been taken to improve the construction codes and monitor the earthquake areas since the Bhuj earthquake?

Sources cited:

1. Bhargava, R.N., V. Rajaram, Keith Olson, and Lynn Tiede, 2016, Ecology and Environment, New Delhi, TERI.
2. CCJM, CDF, and University of Illinois, Chicago, 2019, Water-Based Green Infrastructure – Applications and Guidelines, 30 pp.
3. Wallace-Wells, David, 2019, The Uninhabitable Earth: Life After Warming, Crown Publishing, A division of Penguin Random House LLC, NY, NY.
4. Wikipedia.org, 2018, Tropical Cyclones.
5. Figure 7.1 Courtesy: https://www.researchgate.net/figure/Map-showing-major-river-basins-in-India_fig1_316456071
6. Figure 7.2 Courtesy: https://www.india.com/wp-content/uploads/2016/06/monsoon-maps.jpg
7. Figure 7.7 Courtesy: https://en.wikipedia.org/wiki/Earthquake_zones_of_India#/media/File:India_earthquake_zone_map_en.svg
8. Figure 7.8 Courtesy: https://ncrmp.gov.in/wp-content/uploads/2013/06/Map-showing-Coastal-Areas-of-India-affected-by-Cyclones.jpg

Chapter 8

Environmental Crisis and Social Response

Introduction

Environmental degradation is occurring in all parts of India, especially in heavily crowded cities such as New Delhi, Mumbai and Kolkata. Air is being polluted with burning of biomass and exhaust from industrial/ transportation sources, water is being polluted from industrial and municipal discharges of wastewater into water bodies, and the land is being degraded from improper disposal of solid and hazardous wastes. Although increasing enforcement of industry by pollution control agencies is helping the situation, uncontrolled discharges to the air, water and land are continuing in many urban areas. Unplanned growth of communities within and in the outskirts of urban areas is putting a lot of stress on air, water and land. This is leading to crises related to health impacts and loss of productivity. The air pollution is especially harmful to children and those suffering respiratory diseases. People are wearing masks while traveling in the cities. This chapter will consider the significant impacts of air pollution, water pollution, and land pollution in all parts of India. The social response needed from civil society groups and citizens will also be described. Strong reactions from citizens and civil society groups on a consistent basis are required to tackle the environmental crises in India.

8.1 Air Pollution

Air is required for sustenance of life on this planet. Air is 99.9% nitrogen, oxygen, water vapor and inert gases. Our lungs are constantly taking in air and exhaling the carbon dioxide and other gases. We suffocate when we cannot breathe and when this lasts, we die. The lungs of babies and young adults are very susceptible to impurities in the air and we see increases of asthma and other respiratory diseases when the air quality deteriorates. For those having chronic diseases affecting the lungs, air pollution creates much distress and harm to the lungs. So, it is vital that our governments provide clean air and restrict the pollutants released into the air. This section will deal with the causes of air pollution and how they can be mitigated. It will also briefly describe the laws available to minimize air pollution and how they are implemented.

Air pollution consists of primary and secondary pollutants. Primary pollutants are particulate matter, carbon monoxide, sulfur dioxide, nitrogen oxides, hydrogen fluoride, volatile organic compounds, chlorofluorocarbons, ammonia and aldehydes/organic acids. Secondary pollutants are smog, and ground-level ozone. Smog results from large amounts of coal burning or by vehicular and industrial emissions. In places like New Delhi and Beijing, smog causes deteriorating air quality during several months of the year. Cities have been trying to minimize smog by minimizing diesel truck traffic and using fuels such as compressed natural gas in automobile and truck engines.

Sources of air pollution are many, and can be classified into two main categories: natural and man-made. Natural sources include:

- Dust from large areas of land with little or no vegetation;
- Poisonous gases such as sulfur dioxide, hydrogen sulfide, and carbon monoxide released in the atmosphere from volcanic eruptions;
- Methane emitted by the digestion of food by animals such as cattle;
- Smoke and carbon monoxide from natural forest fires; and
- Carbon dioxide and methane produced from marsh gas production, and biodegradation of organic wastes.

Man-made sources include:

- Deforestation, which results in accumulation of carbon dioxide in the atmosphere;
- Combustion of fossil fuels, which release carbon dioxide and many other harmful gases;
- Transportation, which results in the release of carbon monoxide and nitrogen oxides;
- Industrial activities, which release several harmful gases into the atmosphere; and
- Agricultural activities, which release ammonia and pesticides into the atmosphere.

Table 8.1 details the sources of air pollutants and their effect on living things.

Table 8.1: *Sources of air pollutants and their effect on living things*

Pollutants	*Major sources*	*Typical effects*
Carbon monoxide (CO)	Incomplete combustion of fuels, automobile exhaust, jet engine emissions, blast furnaces, mines, and tobacco smoking	Toxicity, blood poisoning, increased proneness to accidents, and central nervous system impairment
Sulphur dioxide (SO_2)	Combustion of coal and petroleum products, burning of refuse, petroleum industry, oil refining, power houses, sulphuric acid plants, metallurgical operations, and domestic burning of fuels	Increased breathing rate and feeling of air starvation, suffocation, aggravation of asthma and chronic bronchitis, impairment of pulmonary functions, respiratory irritation, sensory irritation, irritation of throat and eyes
Oxides of nitrogen (NO_x)	Automobile exhausts, coal-fired and gas-fired furnaces, boilers, power stations, explosive industry, fertilizer industry, manufacture of HNO_3, combustion of wood and refuse	Respiratory irritation, headache, bronchitis, pulmonary emphysema, impairment of lungs, lachrymatory effect, loss of appetite, corrosion of teeth
Hydrogen sulphide (H_2S)	Coke ovens, kraft paper mills, petroleum industry, oil refining, viscose, rayon manufacturing plants, manufacture of dyes, tanning industry, and sewage treatment plants	Headaches, conjunctivitis, sleeplessness, pain in the eyes, irritation of respiratory tract, respiratory paralysis, asphyxiation, malodorous; blockage of oxygen transfer, poisoning cell enzymes, and damaging nerve tissues in the case of high concentrations
Chlorine (Cl_2)	Accidental breakage of chlorine cylinders, electrolysis of brine, bleaching of cotton pulp, and other process industries using chlorine	Irritation to eyes, nose, and throat, toxicity, respiratory irritation, lachrymatory effects; oedema, pneumonitis, emphysema, and bronchitis in large doses
Hydrogen fluoride (HF)	Glass fibre manufacture, chemical industry, fertilizer industry, aluminium industry, ceramic industry, phosphate rock processing	Irritation, respiratory diseases, fluorosis of bones, mottling of teeth
Carbon dioxide (CO_2)	Combustion of fuels, automobile exhausts, jet engine emissions	Toxic in large quantities, hypoxia
Hydrocarbons (HC)	Organic chemical industries, petroleum refineries, automobile exhausts, rubber manufacture	Some hydrocarbons have carcinogenic effects, lachrymatory effect

Pollutants	*Major sources*	*Typical effects*
Oxidants (such as O_3)	Photochemical reactions in atmosphere involving organic materials and NO_2, reactions induced by silent electrical discharge and intense ultraviolet radiations in the atmosphere	Irritation of lungs, eyes, and respiratory tract; accumulation of fluids in lungs and damage to lung capillaries. These biochemical effects of O_3 mostly arise from generation of free radicals, which attack the SH groups present in enzymes
Dust	Mining activities, asbestos factories, power stations, metallurgical industries, ceramic industry, factory stacks, glass industry, cement industry, foundries	Respiratory diseases, toxicity from metallic dust, silicosis and asbestosis from specific dusts. Asbestos dust causes pulmonary fibrosis, pleural calcification, and lung cancer
Ammonia (NH_3)	Chemical industries, coke oven refineries, stockyards, fuel incineration	Damage to respiratory tracts and eyes, corrosive to mucous membranes
Formaldehyde (HCHO)	Waste incineration, automobile exhausts, combustion of fuels, photochemical reactions	Irritation to eyes, skin, and respiratory tracts
Arsenic (As)	Arsenic containing fungicides, pesticides, and herbicides, metal smelters, by-product of mining activities, chemical wastes	Inhalation, ingestion, or absorption through skin can cause mild bronchitis, nasal irritation, or dermatitis. Carcinogenic activity is also suspected. Attack SH groups of enzymes, coagulate proteins
Cadmium (Cd)	Cadmium-producing industries, electroplating, welding; by-products from refining of Pb, Zn, and Cu, fertilizer industry, pesticide manufacture, cadmium-nickel batteries, nuclear fission plants, production of TEL (tetraethyl lead) used as the additive in petrol	Inhalation of fumes and vapours causes kidney damage, bronchitis, gastric and intestinal disorders, cancer, disorder of heart, liver, and brain, chronic and acute poisoning. Renal dysfunction, anaemia, hypertension, bone marrow disorder, and cancer
Chromium (Cr)	Metallurgical and chemical industries, processes using chromate compounds, cement and asbestos units	Toxic to body tissues, cause irritation, dermatitis, ulceration of skin, perforation of nasal septum. Carcinogenic action suspected
Lead (Pb)	Automobile emission, lead smelters, burning of coal or oil, lead arsenate pesticides, smoking, mining, and plumbing	Absorption through gastrointestinal and respiratory tract and deposition in mucous membranes, cause liver and kidney damage, gastrointestinal damage, mental retardation in children, abnormalities in fertility and pregnancy

Pollutants	Major sources	Typical effects
Zinc (Zn)	Zinc refineries, galvanizing processes, brass manufacture, metal plating, plumbing	Zinc fumes have corrosive effects on skin and can cause irritation and damage mucous membranes
Manganese (Mn)	Ferromanganese production, organomanganese fuel additives, welding rods, incineration of manganese-containing substances	Poisoning of the central nervous system, absorption, ingestion, inhalation or skin contact may cause manganic pneumonia.
Nickel (Ni)	Metallurgical industries using nickel, combustion of fuels containing nickel additives, burning of coal and oil, electroplating using nickel salts, incineration of nickel-containing substances, vanaspati manufacture	Respiratory disorders, dermatitis, cancer of lungs and sinus
Mercury (Hg)	Mining and refining of mercury, organic mercurials used in pesticides, laboratories using mercury	Inhalation of mercury vapours may cause toxic effects and protoplasmic poisoning. Organic mercurials are highly toxic and may cause irreversible damage to nervous system and brain

Source: Bhargava, R.N. et al, 2016

Indoor air pollution from cooking (in developing countries where wood is the fuel) and smoking are deteriorating the health of women and children. Modern household products such as cleaners, paints, disinfectants, and insecticides create indoor pollution if improperly used. In modern office buildings, improper ventilation creates poor air quality which can adversely affect health, and a new term "sick building syndrome" has been created. We have to be aware of these kinds of indoor pollution and minimize their impacts by proper education and monitoring of indoor air quality. Women in developing countries have to be given access to clean biogas instead of the wood fuel that they have traditionally used.

8.2 Water Pollution

Water is essential for life, a person can die without water within 4 days. About half of the world population does not have access to clean drinking water. The range of water borne diseases is significant and these diseases significantly impact the productivity of the population affected. In Africa, China, India, and many countries in Central and South America, water pollution is rampant and is caused by untreated municipal waste discharges to water bodies and by industrial pollution. Technologies are available to treat these wastewaters but in many cases, the funding is not available or misused by government and industry. Although environmental laws and regulations are available in most countries, lack of strict enforcement causes water pollution to increase in poor communities.

Water is needed by people and animals for survival, and is required for agriculture, power production and many industrial uses. If integrated water management were practiced, where water and wastewater are considered as one resource and efforts made to recycle treated wastewater for all our uses, we can have adequate clean water for all. Causes of water pollution include the following:

- Agricultural runoff, which includes ammonia, pesticides and other chemicals;
- Industrial toxic effluents, which include metals and organic/inorganic chemicals;
- Municipal effluents, which include untreated and partially treated sewage; and
- Solid waste that ends up in our water bodies and harms the fish and other creatures that live in the water.

Laws have been passed by the Indian parliament, which generally follow international norms and are designed to prevent water pollution. The enforcement of such laws is in the hands of the State Pollution Control Boards (SPCBs), Central Pollution Control Board (CPCB), and the green tribunals set up by courts of the central government and state governments. The CPCB identified severely polluted stretches on 18 major rivers in India. These stretches were found in and around large urban areas where industrial and municipal effluents are discharged to the rivers. Another major cause of concern is the pollution of groundwater resources from the improper disposal of solid and hazardous wastes. The quantity and quality of groundwater is deteriorating in many parts of the country. Farmers are digging deeper to get the water they need for agriculture and might be pulling up contaminants unknowingly.

The control of water pollution entails the reduction in pollutant loads from human activities discharged to our water bodies so the natural regenerative capacity of the water body is not exceeded. In many cities in India, this is not happening and our water bodies, including major rivers are polluted. This affects the health of many poor people who rely on water bodies, including rivers, for their drinking water needs. By controlling the pollution of rivers and water bodies, we will improve aquatic life, biodiversity and the health of millions of poor people. The ways to reduce this pollution are many and include:

- Treat our wastewater to remove suspended solids and pathogen contamination before discharging to our water bodies;
- Treat our industrial effluents from small, medium scale and large industries. Large and many medium scale industries are treating their effluents but many small companies do not have the capacity or technical knowledge to treat their waste streams;
- Control agricultural runoff into our water bodies so that ammonia and pesticides do not contaminate our water bodies; and
- Manage solid waste in our communities properly so that a large part of it does not end up in our water bodies.

Most water pollutants are eventually carried by rivers into our oceans. In some regions of the world, studies have traced the influence of pollutants up to 100 miles from the river mouth. Concentration of pollutants in the oceans can adversely affect aquatic flora and fauna. Solid waste ending up in the oceans, especially plastics, have a large adverse impact on marine mammals. The suspended solids discharged to oceans may cause turbidity. block light to plants and corals, and clog the gills of some fish species. Unless we address the health of our rivers by preventing pollution, we will leave a deteriorating planet to our children and grandchildren. In addition to all this, global warming is causing damage to many coral reefs around the world. The main types of pollutants are biological, chemicals from agriculture, and industrial toxic pollutants. The two sources of pollution are point sources from industrial and municipal effluents and non-point sources reaching the water bodies from sources such as agriculture that are spread over large areas.

Figure 8.1: ***Point Source pollution.***

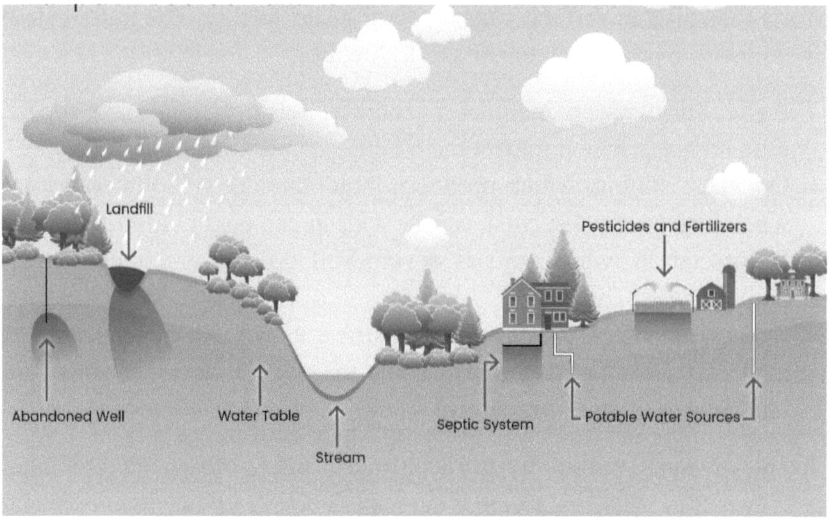

Figure 8.2: ***Non-point source pollution.***
Source: https://www.bcwater.org/threats/

In addition, storm water pollution from parking lots, highways and urban runoff has an adverse impact on the water quality in our rivers and lakes.

Pollution control measures are being undertaken by industry and municipalities around the world. The discharges from these sources are monitored by the State Pollution Control Boards and the Central Pollution Control Board to ensure that the treatment of the waste stream is adequate. However, strict enforcement by these government bodies is lacking and resulting in on-going pollution of our water bodies. The types of wastewater treatment include:

1. Primary treatment to remove the solids;
2. Secondary treatment to remove biological contaminants such as pathogens. Aeration is the main technique to achieve this treatment;
3. Flocculation to remove the suspended solids in water; and
4. Filtration and disinfection to make the water suitable for drinking.

8.3 Land Pollution

Municipal solid waste is a serious problem throughout India. Garbage is found on many city streets and people have not taken seriously the messages from the Central and Local governments to dispose of waste in dustbins that are colored red for dry waste and green for wet waste. In addition, industrial and medical solid waste that is disposed improperly affects the soil and surface water/ground water quality. The main causes for solid waste pollution include the following:

- Agriculture, which includes pesticides, fertilizers, and improperly drained fields which impact the quality of the soil and groundwater;
- Opencast mining, which produces large amounts of solid waste and when improperly disposed, causes widespread land pollution;
- Deforestation, which causes severe soil erosion and making land unproductive; and
- Solid wastes from homes and industries, which is a major source of land pollution. Solid waste landfills make large acres of good land unavailable to other productive uses.

Effects of land pollution on health are many, from contaminated groundwater to soils containing toxic materials that are taken up by plants and are consumed by humans and animals. Toxic gases and odor from landfills impact many living close to the landfills. Figure 9.3 shows the land degradation resulting from opencast mining, mineral dressing and other related activities.

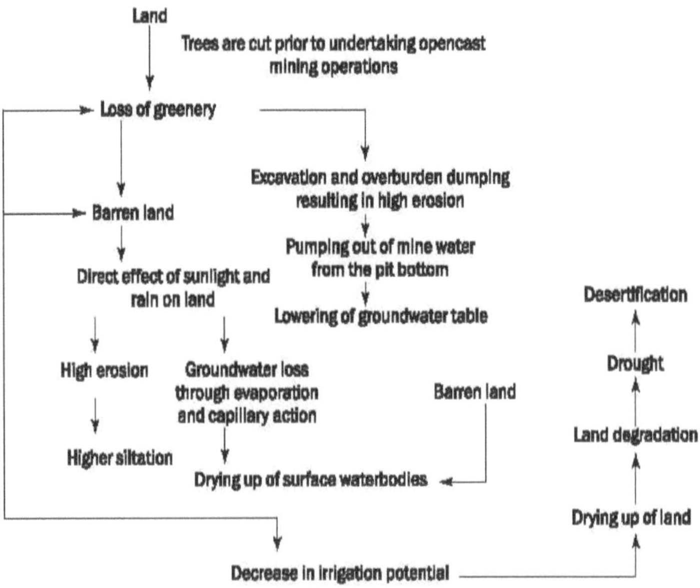

Figure 8.3: ***Land degradation cycle.***
Source: Bhargava, R.N. et al, 2016

Prevention of land pollution is important to maintain agricultural productivity and the surface and ground water quality in our communities.

It starts with the 3 R's, recycle, reuse and reduce. If we use biodegradable plastics, use biofertilizers and natural pesticides in agriculture, and practice sustainable mining techniques (Rajaram et al, 2005), we can significantly decrease land pollution. Other methods of treating and controlling land pollution and degradation include:

- Afforestation, to prevent soil erosion and reduce greenhouse gas emissions;
- Enforcing the environmental regulations at the local, State and Central government agencies;
- Treatment of solid and liquid hazardous waste and sludges to reduce their toxicity and recover useful products before land disposal;
- Implementing biological treatment techniques for municipal solid waste to convert them into compost and biogas;
- Implementing thermal treatment technologies for municipal and biomedical waste to reduce the volume and toxicity of the waste and produce energy and vitrified glass, which is useful for road construction; and
- Recycling construction and demolition waste to the maximum extent possible so that land disposal of solid waste is minimized.

Figure 8.4 provides the options for energy production from municipal solid waste.

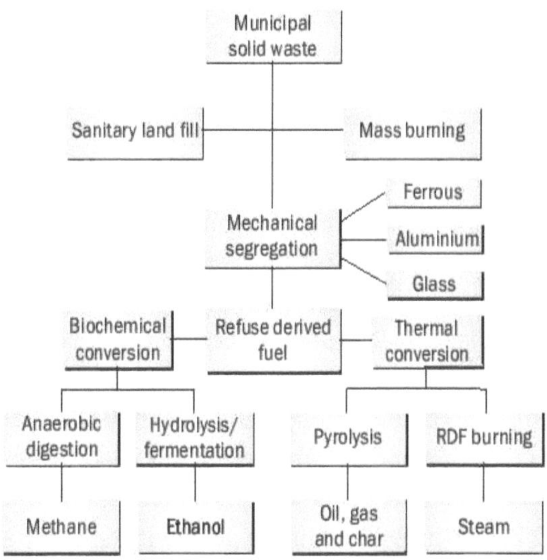

Figure 8.4: ***Options for energy production from municipal solid waste.***
Source: Bhargava, R.N. et al, 2016

8.4 Social Response to Pollution and Status of Environmental Awareness

Society consists of ordinary citizens, professionals and government officials who all have a responsibility to control pollution and create a safe environment in which to live. The U.N.'s Sustainable Development goals have been a worldwide catalyst for education and awareness, forcing countries and businesses to examine environmental practices and proactively work toward a healthy environment on all levels.

The Chipko movement in India was one of the first responses from citizenry to the destruction of forests. People just hugged the trees and would not make way for logging equipment that was being brought to the forests to cut them for timber. Scientific reports by scientists working for the government and universities led to the creation of the U.S. Environmental Protection Agency. Of these, Silent Spring by Rachel Carson, a Marine Biologist working in the US Government, is the most well-known catalyst for the passage of environmental laws in the United States and around the world. The action by Civil Society groups in India resulted in the formation of the Ministry of Environment, Forests and Climate Change, and several environmental laws to protect the Environment.

India has passed several environmental laws to protect the air, water and land resources. Of these, the most impactful include:

- Environmental Protection Act, 1986;
- Air (Prevention and Control of Pollution) Act, 1981;
- Water (Prevention and Control of Pollution) Act, 1974;
- Wildlife Protection Act, 1972; and
- Forest Conservation Act, 1980.

Many other laws relating to the environment and natural resources are provided in Chapter 3, Natural Resources.

Non-governmental organizations (NGOs) play a big part in responding to environmental crises and remaining active to prevent these crises. In addition, the Indian government has passed laws to promote Corporate Social Responsibility (CSR) among India's corporations that have more than 50 employees. Large industrial associations such as the Confederation of Indian Industry (CII) and Federation of Indian Chambers of Commerce and Industry (FICCI) provide training in environmental stewardship to their members and others. CII was formed in 1895 and FICCI in 1927, and there are many other regional industrial associations that promote industry as well as a clean environment. The major NGOs in the Environmental sector are The Energy Research Institute (TERI) and the Center for the Science and the Environment (CSE), both headquartered in New Delhi. There are many regional NGOs doing a lot of good work in environmental stewardship.

The citizen's response to environmental crises is taken up by many lawyers through the Public Interest Litigation (PIL). Improvements in many environmental issues have been accomplished through PILs filed with the Supreme Court of India. The Right to Information Act (RTI) has given ordinary citizens the power to demand action by government agencies and if the agency does not respond within 30 days to a written request, the government agency or concerned official is held responsible. Increased environmental awareness in schools and action by individuals, businesses, and civic groups is improving the environment in many cities. However, the challenge of managing environmental crises remains since the environmental infrastructure to clean wastewater is not adequate and the enforcement of environmental laws is lax.

The Center for Science and the Environment has worked on the issue of air pollution via monitoring, reporting, and leading policy shifts since 1996. Their research and extensive articles provide up to date information on the crisis. Several initiatives highlight the increasing awareness of citizens and efforts to network to put pressure and improve education around pollution. The #Let me Breathe campaign started by journalist Tamseel Hussain, allows people to connect and discuss the increasing problems of air pollution in India. According to an article by Georina Smith for UN Environment, Young Champions of the Environment, "Stories about ordinary folk have reached an estimated 7 million people since the platform launched in October 2017, engaging 20,000 people monthly on Facebook." The Healthy Air Coalition in Bengaluru installed 40 monitors to collect data about air pollution in this city in response to the government's lack of monitoring of the problem. The belief of this group is that with quality data they can effectively drive efforts for change.

The Smart Campus Cloud Network (www.sccnhub.com) has set up a worldwide network of schools and college campuses to combat the environmental effects of climate change. It is helping these schools and colleges improve their campuses to minimize their carbon footprint, and also helping them work on the following Sustainable Development Goals (SDGs) established by the United Nations in 2016:

1. No poverty;
2. Zero hunger;
3. Good health and wellbeing;
4. Quality education;
5. Gender equality;
6. Clean water and sanitation;
7. Affordable and clean energy;
8. Decent work and economic growth;
9. Industry, innovation and infrastructure;
10. Reduced inequalities;
11. Sustainable cities and communities;
12. Responsible consumption and production;
13. Climate action;
14. Life below water (our oceans and rivers;
15. Life on land;
16. Peace, justice and strong institutions; and
17. Partnership for the goals.

Rajendra Shende, Chairman of SCCN, says that "SCCN seeks to accelerate the 17 SDGs, which build on the Millennium Development Goals sponsored by the United Nations in 2000. His organization, based in Pune, India, has unveiled a "SCCN Tool-Kit" to enable innovations in schools and colleges that will improve the environment. These innovations are in areas such as Internet of Things, Artificial Intelligence, Cloud Networking, Machine to Machine Learning, and Block Chain concepts.

Rajendra Shende and his colleagues have also started Terre Policy Centre (www.terrepolicycentre.com) to provide online courses to students and conduct projects that improve the environment in different parts of India. Their projects include rehabilitation of many areas in Maharashtra through afforestation and tree plantings. Terre does research on policy and governance issues related to sustainable development, and conducts a Teachers Olympiad. They also analyze how some of the traditional environmental practices in communities can be revitalized to improve the environment along with economic development.

8.5 Research, Activities and Exercises

1. Review these websites, articles, and reports to get a sense of the urgency and current status of environmental awareness regarding air pollution globally and in India. What conclusions can you draw about

the need to combat this issue? What solutions can you advocate for or efforts can you get involved in?
Center for Science and the Environment
https://www.cseindia.org/page/air-pollution
https://www.cseindia.org/page/air-quality-and-public-health
World Health Organization
https://www.who.int/airpollution/en/
National Resources Defense Council:
https://www.nrdc.org/sites/default/files/clearing-air-reduce-air-pollution-india-fs.pdf
Vox:
https://www.vox.com/2018/5/8/17316978/india-pollution-levels-air-delhi-health
Financial Times
https://ig.ft.com/india-pollution/

2. Audit one's school or home for the use of chemicals. Conduct surveys or examinations of all chemicals used and research ingredients and their effects on health. This can include foods and air and water quality testing. Lobby for your school or home to change methods of cleaning and food sources.
3. Lung capacity testing. With help of simple devices, students can measure and compare their lung capacity with that of asthmatic students.
4. Landfills and waste disposal sites are large contributors non only to the destruction of green spaces, but also to groundwater and land pollution. As India becomes a more consumer society, waste management issues increase. Research general waste management. Find out how and where your trash goes in your community. What is your locality's Waste Management System? Is your waste dumped in landfills? Is it incinerated? Are there people who recycle your waste before or after general disposal? How do industries dispose of waste? What regulations are there and are companies following them? What waste from Industrialized nations is being dismantled in India and how? Gaining awareness is the first step. A simple project is to audit, inventory, and weigh your trash at home and school. Some teachers have had students carry around their personal trash for a week to generate real awareness.
5. Composting is taking natural plant waste and allowing it to break down naturally and turn back into soil. This practice is ancient and

necessary in rural/agricultural communities. Are any homes, local gardens, schools, or businesses composting in your community? It is estimated that about 60% of most urban waste could be composted, and greatly reduce the need for waste disposal. Investigate whether such a program could be implemented in your school or where you live. If you have the space, start and maintain a compost pile. As what is available in a household and what can be done vary widely across India, try to locate local sources of practical advice and information. Be sure that whatever is done is acceptable, both legally and by neighborhood standards.

6. Quantity Based Users Fee—Having people pay for the amount of garbage they create. This is being implemented in many large cities around the world as a way to create awareness and responsibility for one's own waste. Is this happening in India?
7. Extended Producer Responsibility is a new and growing concept that makes manufacturers responsible for their product and packaging from production to disposal. Electronic industries are leading in this area and many have take-back programs when you want to get rid of an electronic device like a computer. It makes the industries more reflective about how a product is produced in order to make sure its component parts can be recycled when disassembled. E-Waste is shipped to India for dismantling, so this is an important area to research to understand India's place in this cycle.
8. Research and create strategies for waste reduction at home and school. Challenge friends and family to buy less. Use both sides of paper for printing and copying. Always consider how an item will be disposed of before purchasing it. Aim for zero waste and research groups promoting this philosophy. One is found at: http://www.ecocycle.org/zerowaste/. EXNORA and other non-governmental organizations (NGOs) in India are leading the way toward Zero Waste Management in cities and villages.
9. Investigate if there are refurbishing centers, recycling plants, demolition depots, in your community. Could there be more, for other products? Be sure to donate all and any items to these places if you no longer want them.
10. Brainstorm ways that simple products could be re-used at home and school: jars for storage and paper for notepads. Cooking oil can be

turned into biodiesel if there are facilities available. Other ideas can be found on many websites. Research them and find inspiration.

11. Go to the website www.sccnhub.com and learn more about the schools and colleges participating in the Smart Campus Cloud Network. Review the tool-kit, and see how your school can use the tool-kit to create innovative ideas that will improve the environment in your community. Also go to www.terrepolicycenter.com, and review the projects completed by them to improve the environment in India. Write a 5-page report on the tool-kit and how you used it to come up with innovative ideas.

References

- Healthy Air Coalition Bengaluru. http://www.healthyaircoalition.org/. Accessed 10 July 2019.
- Smith, Georgina. "#LetmeBreathe: A Social Response to Pollution in India." United Nations Young Champions of the Earth. Accessed 10 July 2019.
  https://web.unep.org/youngchampions/story/letmebreathe-social-response-pollution-india
- Bhargava, R, N., V. Rajaram, Keith Olson, and Lynn Tiede, 2016, Ecology and Environment, New Delhi: TERI.
- Smart Campus Cloud Network (www.sccnhub.com)
- Terre Policy Centre (www.terrepolicycentre.com)

Chapter 9

Environmental and Habitat Conservation

Introduction

The importance of environmental stewardship and habitat conservation is discussed in this chapter. This is critical to minimizing the impacts of climate change and leaving a better planet for future generations. Environmental stewardship includes activities such as planting trees in barren areas, keeping our water resources free of pollution, building and maintaining green infrastructure that minimizes stress on our natural resources, and preventing the buildup of carbon dioxide and other greenhouse gases in the atmosphere. Habitat conservation includes urban planning, protection of forest resources, and conserving green space in rural and urban areas.

The Indian Parliament has promulgated several environmental laws to protect cities, forests, and conserve green space in rural and urban India. These laws are elaborated in this Chapter. The Central and State Pollution Control Boards enforce these laws, and Courts act as the final arbiter of these laws when they are challenged by private or public litigants. The role of the Courts, Non-Governmental Organizations (NGOs), and civil society organizations in environmental and habitat conservation is described in detail. Young people around the world are taking an active part in demanding environmental and habitat conservation by filing lawsuits against the government to force it to become environmentally responsible. NGOs in India such as the Center for Science and the Environment (CSE), the Climate Reality Project, and Confederation of Indian Industries as well as U.S. based ones such as the Sierra Club, National Resources Defense Council, World Wildlife Fund, and Audubon Society are working to raise awareness on these issues. Groups such as 350.org and the youth led

Sunrise Movement and others working for policy shifts are also described. Exercises and activities to understand these topics are provided at the end of the Chapter.

9.1 Environmental Legislation and Enforcement

The Indian Constitution requires the protection and improvement of the environment and safeguarding the forests and wildlife under Article 48 A. Article 51 A (g) mentions the duty of every citizen to protect and improve the natural environment, including forests, lakes, rivers, and wildlife (Bhargava *et al.*, 2016). Various laws passed by the Indian Parliament for environmental management are mentioned below.

- Prevention of Cruelty to Animals Act, 1960
- Wildlife Protection Act, 1972
- Water (Prevention and Control of Pollution) Act, 1974
- Water Cess Act, 1977
- Water Cess Amendments, 2003
- Air (Prevention and Control of Pollution) Act, 1981
- Environmental Protection Act, 1986
- Hazardous Wastes (Management and Handling) Rules, 1989
- Manufacture, Storage, and Import of Hazardous Chemicals Rules, 1989 and 2000
- Public Liability Insurance Act, 1991
- Prior Environmental Clearance and Notification, 2006
- Chemical Accidents (Emergency Planning, Preparedness, and Response) Rules, 1996
- Biomedical Wastes (Management and Handling) Rules, 1998
- Recycled Plastics Manufacture and Usage Rules, 1999
- Fly Ash Notification, 1999
- Municipal Solid Waste (Management and Handling) Rules, 2000
- Batteries (Management and Handling) Rules, 2001
- Biological Diversity Act, 2002
- National Green Tribunal Act, 2010
- E-Waste (Management and Handling) Rules, 2011

All these laws are intended to protect the people and the environment from actions by industry and other private individuals. Enforcement of these Acts is critical to the health and welfare of the Indian society, and the Government is trying to improve enforcement as they gain more experience in doing this and also as more resources are provided to the Government agencies to enforce them. However, the role of individual citizens (through public litigation in Courts) and the NGOs and Civil Society organizations is important in a democracy to ensure compliance with the rules and regulations. The Courts play a major role in enforcing the laws if the government agencies are lax in their enforcement. The role of Courts and NGOs/Civil Society organizations is described in the next two sections.

9.2 Role of Courts

Courts, in a democracy are independent and are the final arbiter of the constitutionality of laws passed by the Parliament. In India, a law was passed in 2010 establishing The National Green Tribunals (NGTs) throughout India. The law provides for the effective and expeditious disposal of cases relating to environmental protection and conservation of forests and other natural resources including enforcement of any legal right relating to the environment and giving relief and compensation for damages to persons and property for matters connected therewith or incidental thereto. It is inspired by India's Constitutional Provision of Article 21, which assures the citizens of India, the right to a healthy environment. Since the formation of these Tribunals, many State and Central Government Courts have set up Green Benches where environmental matters are given priority and decisions handed quickly. The Tribunal is mandated to make and endeavour for the disposal of applications or appeals finally within 6 months of filing of the same. The Tribunal is the first body of its kind that is required by its statute to apply the "polluter pays" principle and the principle of sustainable development. India is the third country following Australia and New Zealand to have such a system.

Initially, the NGT is proposed to be set up at five places of sittings and will follow circuit procedures for making itself more accessible. New Delhi is the principal place of sitting of the Tribunal and Bhopal, Pune Kolkata and Chennai shall be the other places of sitting of the Tribunal. The Tribunal has original jurisdiction on matters of "substantial question relating to the environment" and "damage to the environment due to specific activity". There is restricted access to an individual only if damage to the environment is substantial. The powers of the Tribunal related to an award are equivalent to Civil Courts. (https://en.wikipedia.org/wiki/National_Green_Tribunal_Act).

The Tribunal shall consist of a full-time chairperson, judicial members and expert members. The minimum number of judicial and expert members prescribed is ten in each category and maximum number is twenty. Another important provision included in the laws is that the chairperson, if found necessary, may invite any person or more persons having specialized knowledge and experience in a particular case before the tribunal to assist the same in that case. This law has been instrumental in expedited hearing of many cases related to the environment and public health.

India has a unique provision for Public Interest Litigation (PIL) where a complaint can be lodged directly in the Supreme Court of India for matters of public health and the environment. Many such PILs have been decided by the Supreme Court in the last few decades for matters affecting air and water quality, worker safety, and environment/public health. When the executive branch of the Central or State government does not enforce an environmental law, the Supreme Court issues a mandate for such enforcement. The introduction of Compressed Natural Gas as a clean fuel in New Delhi and many cities was under such mandates, and it has helped air quality significantly.

Courts in the United States have played an active role in protecting the environment. Organizations like the Natural Resources Defense Council have pro-actively used the Courts at local, state, and federal levels to ensure citizen's rights to clean air and water are protected. Since 1970, this group has been involved in the passage of laws such as the Clean Water and Clean Air Act and makes sure through litigations that corporations and government entities are held accountable. A current case, Juliana vs. the U.S., was initiated by the non-profit organization, Our Children's Trust under the leadership of lead counsel and director Julia Olsen. While this organization is also litigating other state and global cases, this one is the most notable because it features 21 youth plaintiffs who are challenging that, through "the U.S. government's affirmative actions that cause climate change, it has violated the youngest generations' constitutional rights to life, liberty, and property, as well as failed to protect essential public trust resources"(Our Children's Trust). The case has been proceeding through the courts and gaining much media attention despite efforts by the government and fossil fuel industry to get it dismissed. If won, it would be an unprecedented landmark case and force the U.S. government at a federal level to take more action to address climate change.

9.3 Role of Non-Governmental Organizations and Civil Society Organizations

The NGO, Center for Science and Environment (CSE) in New Delhi, has done a tremendous amount of advocacy and research work aimed at improving air and water quality, and in exposing the dangers to the environment and public health caused by industries and governmental agencies. Their science based publications and workshops are very widely disseminated and this has raised awareness of various environmental issues throughout India. They promote the greening of schools and businesses, and in 2016 after the government's mandatory directive for all undergraduate college students to take a six month environmental studies course, CSE has been helping train faculty to teach such courses. They also work with other professionals through their institutes to build capacity and knowledge of environmental advocates.

Another organization that is bringing awareness of climate change and its impacts is the Climate Reality Project, started by Nobel Prize winner Al Gore in the United States. Chapters have been created around the world, and they are working with government and civic organizations to educate the citizens on climate change and change public policy to mitigate the impacts of climate change. Climate Reality India (www.climatereality.org.in) is located in Delhi and works with schools and community organizations to minimize the impacts of climate Change.

The Confederation of Indian Industry is India's premier business association which has initiatives that looks at environmental issues as it relates to India's economy. This organization is unique in that it brings together business and governmental goals while also considering what is best overall for Indian society. Through their Centers of Excellence and especially the (CII-ITC) Center of Excellence for Sustainable Development, they have been promoting and training business leaders on how to help achieve the sustainable development goals, reduce carbon, and engage in green building practices since 1991.

In the U.S., Sierra Club, Greenpeace, the Natural Resources Defense Council, and the Audubon Society are some of the oldest environmental organizations whose work in some way or another seeks to preserve our environment. Another newer prominent organization that emerged in 2008 is 350.org. This organization started by environmentalist, journalist, and author Bill McKibben, works in the U.S. and internationally to address and prevent more destruction from climate change by creating a network of organizations and creative efforts to find solutions. 350 was named after 350 parts per million—the safe concentration of carbon dioxide in the atmosphere which the organization hoped the world would not surpass—but has now been exceeded. They provide well researched information to educate about the issues, organize large scale marches and protests, and network with smaller groups to create strategies for real change to stop the use and investment in fossil fuels and transition the world to renewable energy. One initiative being promoted actively is the passage of the Green New Deal by the U.S. Congress. While still in formative stages, the initial proposal would be a program at the federal level to create green jobs (which would also address income inequality) and transition the U.S. to a fossil fuel economy. The Sunrise Movement is a national youth lead organization that is also supporting the Green New Deal as well as many other initiatives to stop climate change and create a green economy. The Alliance for Climate Education has also created multiple education and youth training initiatives to encourage and equip young people to make the needed changes to address climate change.

In addition to these types of formal non-governmental organizations, individuals have contributed much by their own initiatives to organize and protest which has been made easier through the use of social media, allowing such efforts to rapidly motivate the masses on a global scale. Greta Thornburg from Sweden exemplifies this wave of activism, and at age 15 began weekly protests about government inaction on climate change outside the Swedish parliament building. She has given Technology, Entertainment, Design (TED) talks and been an overall inspiration for youth climate activists and others for her determination and dedication to the cause. Inspired by her, a student led Youth Strike 4 Climate event was held March 15, 2019. Youth attended, with many walking out of school to do so in 123 countries to send governments a clear message that adults have not done enough to address climate change.

Activities and Exercises

1. Research the major Public Litigation cases brought by citizens to the Indian Supreme Court and how they have helped improve the health of workers and the citizens of India. In particular, research the role played by Mr. M.C. Mehta in public litigation for protecting the environment.
2. Analyze the work of the National Green Tribunal Courts in improving the air pollution in Agra, Delhi and other major cities of India. How do

these courts operate to expedite the hearing of environmental pollution cases?

3. List the major publications of the Center for Science and Environment during the period 2000 to the present. How have they improved public health and what kind of training do they provide to help citizens and industry combat air, water and land pollution?
4. Research the work of the Audubon Society and describe in a few pages the work they have done to protect diversity of bird species in India and around the world. Will you be interested in being active in the Audubon Society when you are able to join the organization? Why?
5. How has Greenpeace prevented pollution in various parts of India? Do you agree with the strategies used by Greenpeace to force governments and industry to change their behavior and improve the environment?
6. Will you participate in a march organized by a local NGO to protest the deterioration of the environment in your city? If you will not, why will you stay away from such a protest march? If you will participate, how will you convince your parents that you should participate?
7. What are the major industry organizations in India other than Confederation of Indian Industry? How do they operate to protect the environment while representing the interests of industry? How do they educate their members about the environmental laws of the country?
8. Go to www.climatereality.org.in and write a 5-page report as to how they are educating the citizens of India on the impacts of climate change. Would you or your parents get involved in their work?
9. Research Audubon Society and how they have helped citizens save the habitat for birds and appreciate the beauty of birds. How does Audubon Society work in India? Write a 3-page report on Audubon Society's work and how it has improved our environment.
10. Select two of the laws mentioned in Activity 1 and write a 5-page report about its main goals. How are the laws implemented in your community? How have the laws helped improve the quality of the environment?

References

- Bhargava, R.N., V. Rajaram, Keith Olson, and Lynn Tiede, 2016, Ecology and Environment, New Delhi: TERI
- Our Children's Trust. Accessed 12 July 2019 (https://www.ourchildrenstrust.org/juliana-v-us)

Chapter 10

Environmental Approaches to Managing Climate Change

Introduction

Climate change will be with us for the foreseeable future and till we drastically reduce our greenhouse gas emissions, they will continue to adversely affect millions of people around the world. Property damage, death and injury will be significant in all countries around the world, but mostly in the less developed countries that do not have the financial resources and emergency preparedness required to manage the hurricanes, typhoons, tornados, floods, fire and drought that will result from climate extremes. Air quality will deteriorate in major cities and cause severe health problems, especially for the young children and older people. Land productivity will reduce and cause severe famines and food shortages in countries in Africa and Asia. Temperature increases will result in hotter summers and this will result in deaths in many parts of the world that are not prepared to handle these extremes. It is already being seen in parts of India like Rajasthan, New Delhi and other states where the temperature increases combined with the lack of adequate water resources (mainly replenished by the monsoon rains) are causing severe impacts on the population.

David Wallace-Wells states in his book, that a billion people are at risk for heat stress worldwide. One-third of the world's population will risk deadly heat waves at least 20 days per year. The cruelest impacts will be felt by the least resilient of the world's populations. If global warming results in an increase of 2 degrees Celsius (goal of the Paris Agreement), severe droughts will be felt in the Mediterranean and India. This will result in acute food shortages affecting billions of people. Extreme heat in summers could result in roads melting and train tracks buckling. Wildfires are on the increase in California in the US, in Australia, and in many other countries. He suggests that the following is needed:

1. We have to get serious about implementing the Paris Agreement by 2020;
2. New thinking is needed in the areas of transportation, infrastructure (which uses a lot of concrete, a major factor in GHG emissions), energy and agriculture;
3. Living simpler to demand less from planet earth, contrary to the consumer-driven society we live in;
4. Political and moral leadership worldwide to take the necessary steps from policy and technology-driven solutions
5. Selfless, empathetic thinking to help the poor people around the world who will be drastically impacted by the air pollution, water shortages, and food shortages that will result from global warming;
6. Social and financial support systems to be developed for assisting the nations and populations most affected by climate change; and finally
7. Citizen involvement to protect our forests, planting trees to green our cities, and in electing governments that are responsive to the challenges resulting from impacts of global warming in our communities worldwide.

Reduced consumption of natural resources, land and water conservation, use of technology, planting more trees, and better management techniques are required to combat the effects of climate change. Innovation, sustainable management of resources, and technology transfer on a mass scale among countries is required to mitigate the climate extremes. Government policies, private industry participation, and raising awareness of these issues in the general population are critical for success in mitigation efforts. The Paris Climate Agreement has to be implemented and improvements to the agreement to adjust to changing conditions are required. Better health care, including prevention of disease and managing the new viruses that may emerge due to climate extremes, is required on a world-wide scale. These approaches to managing climate change are discussed in this chapter.

10.1 Paris Agreement

The Paris Agreement was discussed earlier in Chapter 2, and the most important part of the agreement is the proposed partnership between the developed and developing nations.

While the Paris Climate Agreement was signed by 185 of the 197 parties to the Convention, signing the agreement must lead to implementation. The UN is helping to facilitate best practices through dissemination of information, but national, state, and local government as well as individuals have to take the steps to develop actionable projects that will ensure the goals of the Paris Climate Agreement which is to prevent the rise in temperature from pre-industrial levels by no more than 1.5-2 degrees Celsius by 2030. Technology transfer and financing of renewable energy technologies and energy efficiency is very important to managing climate change. Technological advances in solar energy have reduced the cost and efficiency of solar energy and this has promoted the widespread adoption of solar energy in many parts of the world. In addition, government incentives to promote renewable energy play a vital part in the adoption of solar energy by industry and individual energy users.

India and the U.S. committed to continue this type of partnership with the USAID office still leading energy efficiency initiatives in India through its Partnership for Clean Energy project. The U.S. India Clean Energy Finance Initiative is also helping to find financing in both countries to support initiatives to create distributed solar energy throughout India. According to Nitin Sethi in an article in *The Wire*, as of November 2018, India was on track to reach its goals of reducing greenhouse gas emissions and transitioning to non-fossil fuel based electricity sources, but was lagging behind in its goal to re-forest and create a carbon sink.

In the United States, the majority of citizens support the Paris Climate Agreement, but the Trump Administration has stated he will remove the U.S. from the agreement that came into effect during the Obama administration. While this is a process that can't be initiated yet and could be impacted by future presidential elections, this federal level messaging has motivated state and local leaders, individuals, and businesses to symbolically sign on to the agreement and create their own networks and legislation to ensure that their towns and states are meeting the guidelines of the Paris Climate Agreement as a way to address climate change. Climate Mayors was one such initiative and is a network of mayors committed to meeting the goals. Other initiatives include America's Pledge, The U.S. Climate Alliance, We Are Still In, and the American Cities climate Challenge. Former NYC mayor and businessman-philanthropist Michael Bloomberg has also been at the forefront of organizing governors and mayors and individuals to remain committed to the Paris Climate Agreement despite the political messages from Washington. In April 2019, Bloomberg donated $4.5 million in 2018 and $5.5 million to the United Nations Climate Change Secretariat to make up for lapsed commitments under the Trump administration (Green).

10.2 Renewable Energy/Energy Efficiency

The range of renewable energy keeps growing based on research being done at universities and companies worldwide. At present, solar, wind and bioenergy are well established along with several energy storage technologies.

Figure 10.1: ***Solar, wind and bioenergy use in India***

Biodiesel and ethanol from corn are in wide use in many parts of the world, and these could reduce in price with ongoing research. Also, the materials that can be used for bioenergy are many and as more research is done, it can be produced from commonly available grasses and other agricultural wastes. The major push for reduction in emissions can be achieved in these areas:

- Transportation (about 15% of total emissions);
- Industry and construction (about 20% of emissions);
- Land use (about 13%); and
- Building heating and cooling, and cement production (about 13%).

Researchers at the University of Wisconsin (UW) in Madison have been working on displacing some expensive chemicals that have been traditionally produced from crude oil. They feel it is easier from an economic standpoint using smaller biorefineries. Scientists in Madison have developed a straightforward process to break plant material into useful feedstock streams (Perspective, UW College of Engineering, 2019). They say that going from biomass to chemicals is cheaper and easier than using petroleum because you are starting with something that is closer to the end product. Another area of promise is making jet fuel from plant sources. The airline Jet Blue in the U.S. has announced a 10-year deal to purchase 330 million gallons of renewable jet fuel (Perspective, UW College of Engineering, 2019).

Renewable energy research holds a lot of promise, and this is demonstrated by the rapid price in solar power as better solar panels are made at lower cost. Areas of other promising research include:

- Fuel cell technology (Figure 10.2) is advancing significantly, and when combined with other technologies to produce hybrid systems, the efficiency can be increased;

Figure 10.2: ***Fuel Cell plant***

- Energy storage technologies are improving as better and lighter materials are found to make the large batteries needed to store wind and solar power.

Figure 10.3: ***Batteries used in car***

- Biotechnological advances are helping produce liquid and gaseous fuels from plants, algae and organic solid waste and sludge; and
- Small scale hydropower in rural communities.

10.3 Artificial Intelligence

Artificial intelligence (AI) is good at taking large amounts of information, sorting through it and classifying it to "learn to detect and predict patterns" that will be useful in weather forecasting. The New York Times International Edition of May 13, 2019, had a special section on Climate Solutions, which describes how AI can help predict weather better to plan for severe weather. As weather forecasting improves, loss of life and property damage resulting from extreme events is minimized (as seen in developed countries). AI is being used in many areas such as energy efficiency improvements in buildings, and in building autonomous driving vehicles. It is also making consumer goods cheaper and better.

Predicting floods and drought events accurately will allow for better planning by government agencies and minimize impact on humans and animals. Disaster management and response is significantly improved with better weather forecasting of cyclones, hurricanes and other natural phenomena. Wildfires in forested areas create a lot of havoc to the weather and damage to property and lives. The unpredictable nature of the winds spreads the fire and if we could predict the weather and develop technologies to control fires. AI can assist in several areas, including:

- Build better electricity systems;
- Predict forest fires so that they can be managed better;

- Make autonomous cars safer and make them more widely available;
- Monitor agricultural emissions and deforestation;
- Create new low-carbon materials;
- Predict extreme weather events;
- Make transportation more efficient;
- Reduce wasted energy from buildings;
- Geo-engineer a more reflective earth (speculative at this time); and
- Give individuals tools to reduce their carbon footprint.

10.4 Policies and Improved Implementation

Governments play an important role in setting policy at the national, state and local levels. Proper policies based on scientific data will help manage climate change and its impacts. Government subsidies and support of research has helped the private sector develop solar and wind power in a big way over the last two decades. Government policy in Israel has resulted in the desert nation becoming a net exporter of food and water technologies. Israel's policy of treating water as a national resource has helped implement meaningful change in managing water and wastewater in the country. It has helped develop desalination technologies to convert sea water into water for drinking, agriculture and industry. Government of India policy related to Rain Water Harvesting has helped capture the rainwater for use in homes and in recharging the aquifers. The Chinese have become leaders in renewable energy and energy conservation through support from the government in research and development and financing.

Implementation, monitoring and enforcement of the policies are needed for them to be effective in mitigating climate change impacts. The Government of India changed the name of the Ministry of Environment and Forest to include Climate Change to show the importance that this subject deserves in the national dialogue. Industry needs tax and other incentives to develop the renewable energy technologies that will replace fossil fuel energy. The Paris agreement provides a framework for governments to act in a concerted manner to minimize greenhouse gases and mitigate climate change. The Indian Government has provided subsidies for solar and wind energy and is promoting bioenergy through the National Biomethanation Plan. The Ministry of New and Renewable Energy (MNRE) and the Department of Science and Technology have provided leadership in the Research and Development of alternatives to fossil fuels. The MNRE in 2017 achieved 50,000 MW (megawatts) of grid based renewable energy and 1400 MW of off-grid energy (Wikipedia, June 25, 2019). Its main mission consists of:

- Provide energy security;
- Increase production of bio, wind, solar, hydro, geothermal and tidal energy;
- Energy availability and accessibility; and
- Energy affordability and equity.

The United States Congress in 2019 proposed a Green New Deal (similar to the one proposed by President Franklin Roosevelt during the Great Recession of 1930s) to combat climate change and income inequality (The Washington Post, May 5, 2019). The author of the article Ben Adler states that "unless developing countries get financial help to develop sustainably, with industrialization powered by renewable energy, they will use carbon-based energy sources." A $100 billion commitment mentioned by the US in 2009 has not been fulfilled, and as of 2016, the number is $70 billion, according to a United Nations estimate. The Paris agreement has mechanisms for monitoring such assistance and the technology transfer that was promised in the agreement.

New York State and New York City have taken the lead in the U.S. in creating bold legislation to move the state and city rapidly toward a green future. In January 2019, NY State Governor Andrew Cuomo created a Green New Deal for the state which aims to have the state reach 100% clean energy by 2040, create green jobs, and ban fracking and coal powered electricity. New York City signed off on a similar initiative in April 2019 called the Climate Mobilization Act. While some elements of the plan are controversial, the goal is also to reduce energy emissions of buildings through retrofits of windows, heating, cooling, and green roofs, transition NYC to 100% renewable energy while cutting carbon emissions, and create green jobs in the process. It is the largest single mandate to curb climate change of any city in the world.

10.5 Resource Conservation

As the population of the world increases, we will run out of clean water, fertile land for growing food, and clean air in our cities if we do not start conserving natural resources and stop pollution of water, land and air. Lions Club International Foundation added Environmental Protection as an area of support in 2017, and launched a program for tree planting worldwide to fight the effects of Climate Change (Lion Magazine, March 2019). The magazine mentions the following value of trees to communities and the environment:

- Trees make communities healthier – they reduce asthma occurrence, and promote exercise to reduce obesity;
- Trees clean the air we breathe - urban trees capture fine particles from the air as well as greenhouse gases. Value of this in the US is about $7 billion per year;
- Trees save energy – shade from trees reduces the need for air conditioning in summer and provides a wind break in winter.
- Trees shelter and feed wildlife – they provide wildlife habitat and pollen for the bees.
- Trees naturally manage stormwater – the US Department of Agriculture estimates that 100 mature trees intercept about 250,000 gallons of rainfall per year; and
- Trees increase property values – homes in neighborhoods with mature trees increase the value of the homes by 3.5 to 10 percent.

Trees are carbon sinks and afforestation is an effective way of reducing greenhouse gases in the atmosphere. The Paris agreement promotes afforestation in many parts of the world through providing financing for such programs. In addition to preserving and planting more trees, resources can be conserved in the ways listed below.

- Increasing efficiency of water use in our cities and preventing water leaks in our distribution systems;
- Recycling metal, glass and paper waste in our cities and rural areas;
- Composting organic solid and food wastes to produce useful compost;
- Using treated wastewater for agriculture and implementing efficiencies in water use to grow our food;
- Preventing water pollution in our rivers and lakes through proper municipal and industrial waste management;
- Preventing air pollution and accumulation of greenhouse gases in the atmosphere; and
- Efficiency in all aspects of energy use and using renewable energy sources.

Dr. Sailesh Rao, originally a Stanford educated electrical engineer who was involved in the early development of the internet, changed careers and has become an ardent advocate of the transition to veganism and vegetarianism as a solution to climate change. In his thinking, ending human reliance on animal agriculture is the only fast enough way for land to be freed up, reforested and regenerated with plant life to allow the absorption of excess carbon. Through his films such as Cowspiracy, books, and the non-profit Climate Healers, he is ardently sharing his vision of how to solve climate change. While some might consider his solution extreme, he provides thoughtful fact based evidence to support his views on how humanity might heal the planet by accepting that we are the "thermostat species" and "stewards of life."

Green America, a U.S. based non-profit that has been building and coordinating multiple efforts to foster the green economy in the U.S. has started an interesting initiative called the The Carbon Farming Innovation Network through its Center for Sustainability. The goal is to promote and coordinate efforts to engage in regenerative farming techniques which will improve the health of the world's soil. Industrialized heavy pesticide and fertilizer-based farming techniques have depleted the soil, harming its natural ability for food production, impacting human health for both agricultural workers and food consumers, and hindering the soil's ability for carbon sequestration. It is believed that by fostering healthy practices such as no-till cropping, groundcover and cover crop development, agroforestry, and use of mulching, crop-rotation, manure, and composting, soil organic matter can be improved leading to a better biodiversity of the living organisms which allow for better carbon sequestration in the soil. Many other entities are exploring soil regeneration as a viable option for controlling climate change and much inspiration for these efforts has come from Vandana Shiva, creator of the Indian NGO Navdanya. Her recent book Soil or Not outlines the crisis and solutions via improved farming eloquently.

10.6 Technologies for Managing Extreme Weather

Global warming has resulted in extreme weather patterns around the globe. Paris had temperatures of 45 degree C, the highest temperature in the city in June 2019. India has had extreme heat events where the roads have buckled and thousands have been killed. The periods of severe drought, fire, and floods are increasing year by year in the United States. We have to use all the techniques mentioned in this chapter to manage these extreme weather events and reduce the impact on lives and property. Innovation and community problem solving are the only ways to deal with these events, and preventive measures are always better than the costly cures such as building sea walls in our coastal cities (like Venice, Italy, did a few years ago). The human brain is capable of solving any problem if collective action is taken for the common good rather than feeding the individual greed of the few.

Technologies mentioned in this chapter along with wise long-term policies of governments around the world will be able to manage extreme weather with the least amount of property damage and loss of lives. It is up to each individual to do the right thing – conserve water and energy, use less packaged products and support the local economy with purchases of local goods, drive less and walk more, and participate in the betterment of the community.

10.6.1 Biochar

Biochar is a solid material obtained from the thermochemical conversion of biomass in an oxygen-limited environment (Biochar). The International Biochar Initiative is developing centralized, decentralized and mobile systems to produce biochar for various uses. The most common crops used for making biochar include various tree species, and agricultural residues. Residue to Product Ratio (RPR) and the collection factor (CF) the percent of residue not used for other purposes, measure the approximate amount of feedstock that can be used for biochar after harvesting. For example, Brazil harvests approximately 460 million tons of sugarcane annually, with an RPR of 0.30, and a CF of 0.70 for the sugarcane crop, which is usually burned in the field and causes a large amount of air pollution. This is happening in India, where burning of agricultural residues are burned in the field and causes harming air pollution in the largest metropolitan area of Delhi.

The physical and chemical properties of biochar as determined by feedstocks and technologies are crucial for the application of biochar in the industry and environment. The properties of biochar can be characterized in several respects, including the proximate and elemental composition, pH value, porosity, etc., which correlate with different biochar properties. The primary property of biochar that has been used to-date is its ability to store carbon in a stable way in the ground for centuries, potentially reducing or stalling the growth in atmospheric greenhouse gas levels. In addition, biochar can improve water quality, increase soil fertility, raise agricultural productivity, and reduce pressure on old-growth forests. If biochar is used for production of energy rather than as a soil amendment, it can be directly substituted for any application that uses coal. Pyrolysis also may be the most cost-effective way of electricity generation from biomaterial.

Bates and Draper state that "evidence is rapidly mounting that the potential of harnessing carbon to reverse climate change extends far beyond agriculture. We seek to share the growing number of ways that carbon can be used above the ground to improve health, rebuild infrastructure, provide or boost renewable energy production, rebalance atmosphere and ocean equilibrium, and offer a host of other benefits. Practitioners from around the world are demonstrating that carbonizing biomass can help urban and rural areas become more sustainable and regenerative by reducing waste, restoring ecosystems, closing nutrient cycles, and slashing emissions".

The authors provide detailed information and case studies about how biochar can be used for:

- concrete and asphalt for infrastructure,
- improving plastics,
- improving paper products,
- improving water treatment,
- improving our livestock,
- producing bioenergy,
- manufacturing of consumer products, and
- case studies of grassroots solutions for using biochar in rural communities around the world.

Using these applications for biochar, the authors estimate that about 50,6 gigatons of carbon dioxide or equivalents per year. This exceeds the present human-made emissions by 10 gigatons, meaning that in combination and at full scale, they could pull that much carbon dioxide from the atmosphere and oceans annually. We could restore a safe climate even faster if emissions were to decline as steeply as the Paris agreement requires.

Intensive research into manifold aspects involving biochar is underway around the world. From 2005 to 2012, there were 1,038 articles that included the word biochar in the topic that had been indexed in the ISI web of Science (Biochar). Biochar sequesters carbon in soils because of its prolonged residence time, ranging from several years to millennia. In addition, biochar can promote indirect carbon sequestration by increasing crop yield while potentially reducing carbon mineralization. Students at Stevens Institute of Technology in New Jersey are developing supercapacitors that use electrodes made of biochar. Researchers at University of Florida have developed a process that removes phosphate from water using biochar. Pyrolysis of biomass to produce biochar and energy can become cost-effective when the cost of a ton of carbon dioxide reaches $37.

10.6.2 Efficient Water Management

Peter Gleick of Pacific Institute lists five-hundred water related conflicts since 1900, almost half of the entire list is since just 2010. The five-year Syrian drought that stretched from 2006 to 2011, producing crop failures that created political instability and helped usher in the civil war that produced a global refugee crisis, is one vivid example. In India, water sharing conflicts between the states has taken place many times, resulting in destruction of property and lost lives. The cycle of droughts and floods is increasing all over India in the last decade, resulting in loss of thousands of lives, and creating farmer suicides.

David Wallace-Wells states that globally, between 70 and 80 percent of freshwater is used for food production and agriculture, with an additional 10 to 20 percent set aside for industry. He states that an abundant resource is made scarce through government neglect and indifference. bad infrastructure and contamination, careless urbanization and development. Some cities lose more water to leaks than they deliver to homes; even in the United States, leaks and theft account for an estimated loss of 16 percent of freshwater; in Brazil, the estimate is 40 percent. He continues to point out that the global result is that as many as 2.1 billion people around the world do not have access to safe drinking water, and 4.5 billion don't have safely managed water for sanitation (David Wallace-Wells, 2019). However, this acute humanitarian problem can be solved by using technologies and policies developed by Israel over the last 70 years.

Israel is the world leader in water management technologies and Israelis have transformed the Arava desert in the south of Israel to bounty vegetable farms by using water wisely. They are now exporting food and water to their neighbors and providing efficient water management technologies to countries around the world. The use of drip irrigation was developed in the late 1960s in Israel and its use is now widespread in India and many parts of the world. In recent years, they have developed soil moisture sensors that allow one to provide water to a plant only when it needs it, and not waste water that the plant cannot use. Desalination technologies have been perfected in Israel and these have reduced the cost of desalination. Wastewater treatment and recycling technologies have been perfected in this small country. Their water management policies prevent the waste of water during distribution and supply and their water pricing policies have essentially stopped all water wastage in cities and rural areas.

Greywater is the water from all our household uses except the toilet wastewater which is called Blackwater. Simple technologies are available to treat and use the greywater in our toilets and gardens, and should be widely adopted in all countries around the world. This will save our treated water for cooking, bathing/washing and drinking only. By doing this, we can be water efficient and conserve clean water for human and animal consumption. Treated wastewater without pathogens contains nutrients that are critical for agricultural productivity. Using treated wastewater has proven to increase agricultural productivity in the eastern US and also prevent eutrophication of lakes and other water bodies. Innovations in wastewater treatment to make it safe while removing the nutrients for beneficial use are constantly being made. Many large wastewater treatment plants have become energy neutral by producing the energy they need from sludge generated during treatment and supplementing it with organic solid waste from the community. Such practices have been adopted widely in all parts of the world to conserve energy while reclaiming clean water for reuse.

Storm water management technologies have been improving over many decades, and now green infrastructure is available to recover most of the storm water and use it in our cities. Bioswales, water cisterns, and many techniques to store the storm water underground are available to avoid flooding in our urban areas while recovering the storm water for beneficial use. Such technologies should be widely disseminated around the world and implemented to conserve water and prevent the cycle of floods and droughts.

Finally, water conservation measures are the most inexpensive and can be implemented through a change of mindset. Hotels in India now serve water in glass bottles and provide glasses so that customers can use what they need instead of wasting water by taking a few sips from a glass filled with water and wasting the rest. Many such innovative ways to save every drop of water are needed to manage water availability in periods of extreme weather fluctuations. Taking shorter showers, recycling the greywater, improving water use in agriculture and industry, and pricing water in a manner that encourages conservation are simple techniques for assuring water availability in the future.

10.6.3 Recycling Plastics and other Waste Streams

A scientist in Singapore has developed a method to transform plastic waste into energy using a chemical process instead of the pollution-creating incineration method. Soo Han Sen of Nanyang Technological University, Singapore, has developed a carbon-neutral method of breaking down non-biodegradable plastics such as polyethylene into chemicals that can be used to produce energy. Waste streams from construction and demolition can be segregated, processed and recycled for new infrastructure. Many cities are demanding that such recycling technologies be used before developers and contractors get permission to work on a project. The Cradle to Grave policy of the past is being replaced by a cradle to cradle policy which is promoting innovation in the management of waste streams.

Human innovation along with carbon neutral policies can transform the waste generated by society into useful products. Many universities, entrepreneurs, corporations and government laboratories are working on minimizing waste by recycling them for beneficial uses. There is hope for mankind to manage the extreme weather changes and prevent them from happening by adopting appropriate technologies and wise government policies. Survival of humans, especially in vulnerable island nations, demands that we undertake the right steps to prevent catastrophes around the world.

10.7 Activities and Exercises

1. Fully review the words and terms of the Paris Climate Agreement, the UN's Intergovernmental Panel on Climate Change Report, and the Climate Action Tracker. Find individuals, government entities, NGOs, or businesses that are engaging in projects that match with the stated goals to see how the goals of the treaty are being acted upon to address the real needs of society. Compare and contrast goals set with actual steps being taken to identify additional ways to advocate for change.
   Paris Climate Agreement https://unfccc.int/process-and-meetings/the-paris-agreement/the-paris-agreement
   IPPC https://www.ipcc.ch/sr15/
   Climate Action Tracker https://climateactiontracker.org/

2. Read books and watch movies on current efforts to create new technologies and find solutions to solve our environmental issues. Explore Green America's, Navdanya, and the Climate Healers websites to start.
   Green America https://www.greenamerica.org/
   Navdanya http://www.navdanya.org/site/
   Climate Healers http://www.climatehealers.org/
3. Read the National Climate Action Plan on Climate Change published by the Government of India, and document how the Central Government is planning to minimize greenhouse gases and how renewable energies are being implemented in your State. Write a 5-page report with your findings.
4. How is Artificial Intelligence being used in various fields in India? Write a short report on how Artificial Intelligence can be used to monitor the use of energy and water in your home so that you minimize energy and water use.
5. Review the website of your State Pollution Control Board (SPCB), and write a 5-page report on policies and rules being implemented in your state to monitor and reduce air and water pollution. Is your local government implementing the SPCB rules?
6. Imagine you are the Community or Resident Welfare Association Conservation Officer.
   What would you do to conserve energy and water in your community? How would you maximize recycling of solid waste in your building? How would you set up a composting center in your community to convert food waste into compost for the kitchen gardens? Write a 5-page report.
7. Select one of the technologies mentioned in 11.6 and do independent research. Write a 5-page report on the technology and if that technology can be used in your community or state.

References

Bates, Albert and Draper, Kathleen, Burn: Using Fire to cool the Earth, Chelsea Green

Publishing, Vermont and London, 2018.

Biochar, https://en.wikipedia.org/wiki/biochar, Accessed on 6th July 2019.

Government of India, National Plan on Climate Change, Ministry of Environment and Forests and Climate Change, 2012, 52 pp.

Green, Miranda. "Bloomberg Donates $5.5 million to Fill in Paris Agreement Gap." The Hill. 22

April 2019. Accessed 14 July 2019. https://thehill.com/policy/energy-environment/440013-bloomberg-donates-55-million-to-fill-in-paris-agreement-gap.

https://www.greenamerica.org/blog/farmers-and-companies-unite-fight-climate-change-healthy-soil

Sethi, Nitin. "India Set to Achieve 2 of 3 Paris Climate Agreement Goals: Draft Government

Report." The Wire. 12 November 2018. Accessed 14 July 2019 https://thewire.in/environment/india-paris-climate-agreement-targets

Wallace-Wells, David., 2019, The Uninhabitable Earth: Life After Warming, Crown Publishing, NY, 2019.

Chapter 11

Indian and International Case Studies

Introduction

Case studies are an effective way to describe how different projects and movements in India and around the world are making the environment better. They inspire people to get involved and start improving the environment in their community or their country. They also teach us lessons learned from these case studies and these lessons can be used to further improve our environment. Case studies have been chosen from India and Asia, and provide details of how these cases have improved the environment and lessons we have learned from them. The projects included in the case studies include the following:

- Chipko Movement to protect the forest habitat;
- Project Tiger to protect Tiger Habitats
- Swachh Bharat Campaign to create a clean India
- Fukushima nuclear plant disaster in Japan
- Oil Spills and impacts on the environment.

11.1 Chipko Movement

The Chipko movement or Chipko Andolan was to save trees or conservation of forests in India where people prevent trees from being cut down by hugging them and not letting logging equipment come near the trees (https://en.wikipedia.org/wiki/Chipko_movement). It began in April 1973 in Reni village of Chamoli district, Uttarakhand, and went on to become a rallying point for many future environmental movements all over the world. It created a precedent for starting a nonviolent protest in India to save the environment, and its success meant the world immediately took notice of this movement. It inspired many such eco-groups by helping them slow down the rapid deforestation, expose vested interests, increase ecological awareness, and demonstrate the viability of people power. It stirred up the existing civil society in India, which began to address the issues of tribal and marginalized people. The Chipko Andolan practiced methods of Satyagraha, the nonviolent movement started by Mahatma Gandhi to gain independence for India. Both male and female activists from Uttarakhand played vital roles, including Gaura Devi, Suraksha Devi, Bachni Devi and Chandi Prasad Bhatt.

Increasing hardship was faced by the tribal people in Garhwal Himalayas (in Uttarakhand) since they were losing their forests to unscrupulous contractors who logged the trees in their area without compensating them adequately. This started rising ecological awareness of how reckless deforestation had denuded much of the forest cover, resulting in the devastating Alaknanda River floods of July 1970 resulting from a major landslide which blocked the river. Soon, villagers and women began to organize themselves under smaller groups, taking up local causes with the authorities, and standing up against commercial logging operations that threatened their livelihoods. In October 1971, the Sangh (Dasholi Gram Swarajya Sangh) workers held a demonstration in Gopeshwar to protest against the policies of the Forest Department. More rallies and marches were held in late 1972, but to little effect, until a decision to take direct action was taken.

The first such occasion occurred when the Forest Department turned down the Sangh's annual request for ten ash trees for its farm tools workshop, and instead awarded a contract for 300 trees to Simon Company, a sporting goods manufacturer in distant Allahabad, to make tennis rackets. In March 1973, the lumbermen arrived in Gopeshwar, and after a couple of weeks, they were confronted at the village Mandal on 24 April 1973, where about a hundred villagers and Sangha workers were beating drums and shouting slogans, thus forcing the contractors and the lumbermen to retreat. This was the first confrontation of the movement. The contract was eventually cancelled and awarded to the Sangha workers instead. The growing concern over commercial logging and the government's forest policy forced Sangh to resort to tree hugging, or Chipko, as a means of nonviolent protest.

But the struggle was far from over, as the same company was awarded more ash trees, in the Phata forest, 80 km away from Gopeshwar. Here again, due to local opposition, starting on 20 June 1973, the contractors retreated after a stand-off that lasted a few days. Thereafter, the villagers of Phata and Tarsali formed a vigil group and watched over the trees until December. The lumbermen retreated leaving behind the five trees felled. The final flash point began a few months later, when the government announced an auction scheduled in January 1974, for 2500 trees near Reni village, overlooking the Alaknanda River. Bhatt set out for the villages in the Reni area, and incited the villagers, who decided to protest against the actions of the government by hugging the trees. Over the next few weeks, rallies and meetings continued in the Reni area.

On 25 March 1974, the day the lumberman were to cut the trees, the men of Reni village and Sangh workers were in Chamoli, diverted by State government and contractors to a fictional compensation payment site, while back home labourers arrived by the truckload to start logging operations. A local girl, on seeing them, rushed to inform Gaura Devi, the head of the village Mahila Mangal Dal, at Reni village. Gaura Devi led 27 of the village women to the site and confronted the loggers. When all talking failed, and the loggers started to shout and abuse the women, threatening them with guns, the women resorted to hugging the trees to stop them from being felled. This went on into the late hours. The women kept an all-night vigil guarding the trees from the loggers until a few of them relented and left the village. The next day, when the men and leaders returned, the news of the movement spread to neighbouring Laata and other villages including Henwalghatti, and more people joined in. Eventually, after a 4-day standoff, the contractors left.

The news soon reached the State capital where the then Chief Minister, Hemwati Nandan Bahuguna, set up a committee to look into the matter, which eventually ruled in favour of the villagers. This became a turning point in the history of the eco-development struggles in the region and around the world. The struggle soon spread across many parts of the region, and such spontaneous stand-offs between the local community and timber merchants occurred at several locations, with hill women demonstrating their new-found power as non-violent activists. As the movement gathered shape under its leaders, the name Chipko movement was attached to their activities. The word originally coined by Bhatt was "angalwaltha" in the Garhwali language for embrace, which later was adapted to the Hindi word, "Chipko", which means to stick. Over the next five years, the movement spread to many districts of the region, and within a decade, the whole of Uttarakhand Himalayas. The villagers demanded that no forest exploiting contracts should be given to outsiders and local communities should have effective control over natural resources like land, water and forests.

The movement achieved a victory when the Prime Minister of India issued a ban in 1980 on felling trees in the Himalayan regions for fifteen years, until the green cover was fully restored. One of the prominent Chipko leaders, Gandhian Sunderlal Bahuguna, took a 5,000 kilometers trans-Himalayan foot march in 1981-1983, spreading the Chipko message to a far greater area. Gradually, women set up cooperatives to guard local forests, and also organized fodder production at rates conducive to the local environment. One of the Chipko's most salient features was the mass participation of female villagers. Environmental activism was born in India and it inspired women all over the world to fight for environmental protection. The movement also inspired many lawyers in India to file Public Interest Litigation to force the government to pass environmental laws and regulations to protect the health of poor people and protect the environment.

Figure 11.1: ***Origin areas of Chipko Movement.***
Image Courtesy: https://www.khub.info/chap_practice.php?id=27&pid=12

11.2 Fukushima Nuclear Disaster

Before we discuss the Fukushima nuclear disaster, a background on nuclear radiation and nuclear power is necessary to understand this section. As India is a nuclear power sandwiched between China and Pakistan which also have nuclear weapons, every Indian should have knowledge to dispel the mysteries of the atom. With peacetime uses of nuclear energy and medical use of radiation proliferating, the need is even greater. Figure 12.1 shows solar radiation and how it impacts the earth.

Like most forms of pollution, radiation comes in many strengths and forms, some toxic, some harmless, some avoidable and some inescapable. Unlike most types of pollution, radiation cannot be detected without special instruments. Not much more than a century ago, there was no human caused radiation, little knowledge of the kinds of radiation and even less of the dangers of some forms. There were no medical uses of radiation, no nuclear power plants or nuclear weapons, but radiation was there. Then and now, people were bathed in radiation, background radiation, and it is against that background that current concerns about radiation must begin.

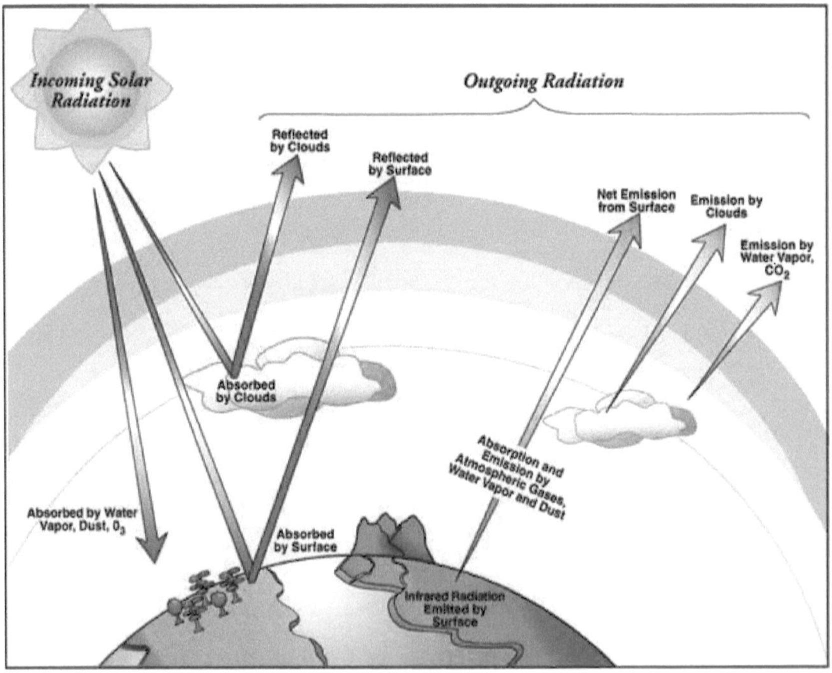

Figure 11.3: ***Solar Radiation.***

Thus, background radiation will vary from location to location, typically lower at sea level than in the mountains. Lower altitudes are no guarantee of lesser amounts of background radiation, however, as some minerals contain radioactive isotopes. There is radon in soil and water, coming from breakdown of naturally occurring uranium and concentrating in enclosed spaces in buildings, especially in lower levels. [see Figure 12.4] As a gas, radon can be inhaled, releasing its radiation in the lungs and is a likely cause of lung cancer among people who have never smoked, as well as multiplying the risk among those who do.

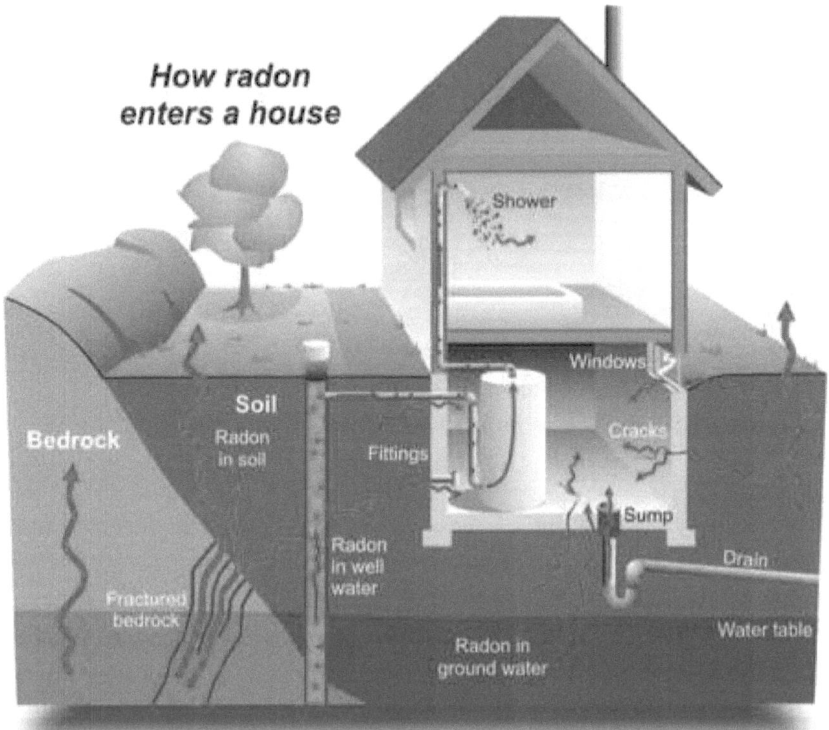

Figure 11.4: ***Radon in Soil & Water.***

In the modern era, there is additional exposure to radiation, from medical procedures including x-rays, from nuclear power, from residual levels from atmospheric weapons and even from industrial use. At current levels of consumption, more radioactive material may be getting into the environment from burning coal in power plants than from nuclear power plants as coal contains small amounts of uranium and other radioactive isotopes. [see Figure 11.5]

Figure 11.5: ***A nuclear power station with steam coming from the cooling towers.***

Historically, measurement of radiation involved several different units depending on the type of radiation and what was being measured: the energy of the radiation, the number of radioactive decays or the effect on humans. As the effect of the radiation is of most concern, the unit of effect called the Sievert (abbreviated Sv) is used, although the older unit the rem (1 Sievert = 100 rems) will often be seen. Rem stands for radiation equivalent man. An average background level is about 2.5 mSv, that is 2.5 milli-Sievert or 0.25 rem though the averages vary widely. In some areas of Kerala near the coast, measurements average three times as much but can go as much as 30 times higher.

When the early investigators of radioactivity directed the stream of radiation from radium or other source past a magnet, they quickly found that some bent one way, some the opposite and some not at all (See Figure 11.6). They called the three main types of radiation alpha (α), beta (β)and gamma (γ or Y) radiation after the first three letters of the Greek alphabet.

Penetration through metal foils also distinguished the alpha from the beta, with the beta penetrating about 100 times as much as the alpha, and the gamma far more. While there are other types of radiation from numerous subatomic particles, the original three, for some purposes with neutrons added, are sufficient for describing health effects.

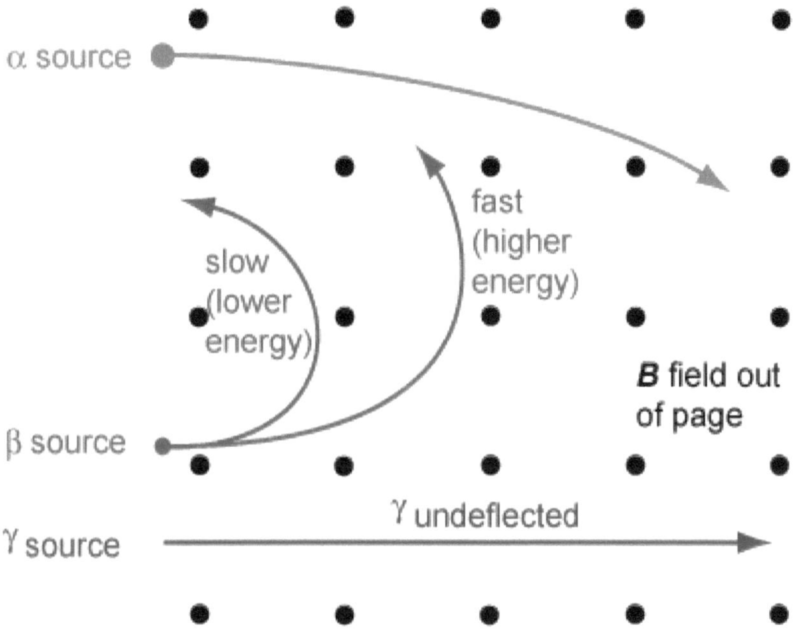

Figure 11.6: ***Effect of magnetic field on α, β & γ rays.***

All three can be damaging to living cells, but the amount of damage depends on the location of the source of the radiation. Both alpha and beta radiation are stopped or shielded by a small thickness of common materials, even paper. As a gas, radon gets into the lungs. Radon is an alpha emitter as is its decay product, polonium, and even in fairly low concentration in a home may have the effect in a year of multiple x-rays. Gamma radiation is much more penetrating, with even substantial thickness of any material water, concrete or lead, not giving complete blocking and thus is the major concern of external radiation.

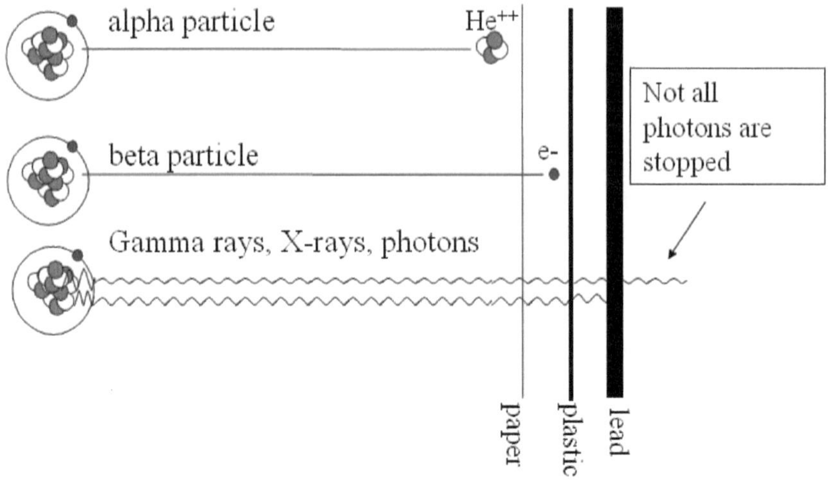

Figure 11.7: ***α, β & γ rays.***

Of course, radiation is valuable in medicine, for diagnosis and treatment. The side effects of exposure to ionizing radiation for medical purposes are the same in nuclear accidents, including nausea, but can be monitored better. More localized radiation, directed at a cancerous tumor, has far lesser bad effects while destroying the tumor. Radiation and radioactive isotopes are also useful in diagnosis and treatment of cancers. Like x-rays, the harm done by radiation has to be balanced against the value of diagnosis and treatment.

Radiation linked to nuclear energy and weapons contrasts with medical uses of radiation. Medicine uses many different isotopes; however, nuclear power and weapons use only a select few, primarily U-235 and Pu-239. All radioactive isotopes give off energy when they decay, as well as particles, alpha (α) and beta (β) and others. Both uranium and plutonium undergo fission, splitting into roughly equal parts. When U-235 splits, one possible pair of the elements produced (there are several) is barium and krypton. The splitting is initiated by a neutron, and the process itself produces excess neutrons. Those neutrons can go on to impact more uranium atoms, increasing the amount of fission in a chain reaction. [see figure 12.8] However, the amount of uranium must be right: too large and the reaction would be spontaneous, out of control; too small and the chain reaction would fizzle as too many neutrons would escape without hitting another uranium atom. The amount that is right for the reaction to be self-sustaining is called the critical mass.

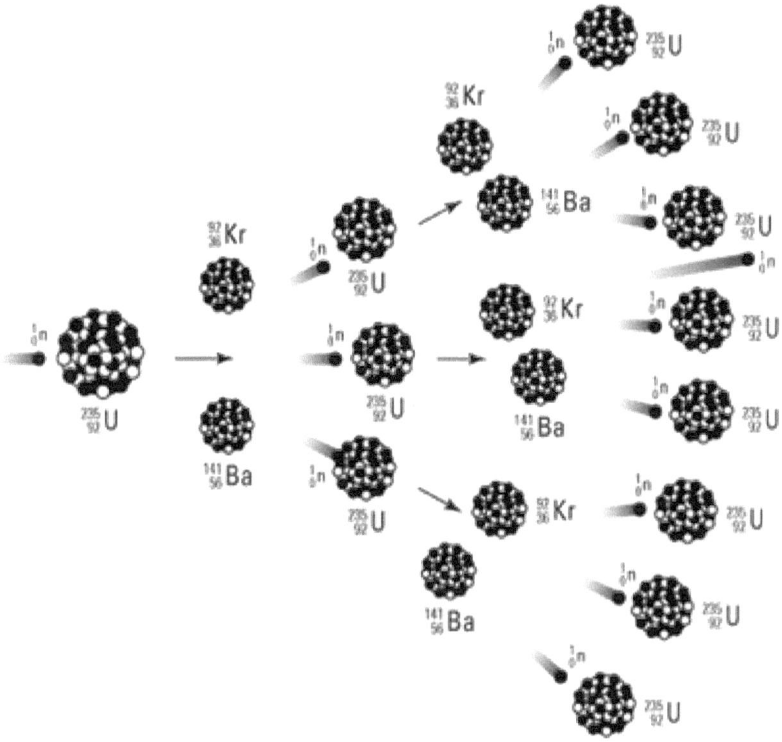

Figure 11.8: ***Critical mass of U-235 splits.***

In an atomic bomb, two subcritical masses are kept separate until detonation is intended. In a power plant, control rods keep the flux of neutrons stable, preventing the runaway reaction. Also, in a power plant, the percent of U-235 is low, about 3%, well below the 90% needed for a bomb. Consequently, the uranium in a nuclear power plant cannot explode like a bomb though, as happened at the Chernobyl nuclear disaster in Ukraine, the reactor can have an uncontrolled chain reaction.

While the devastating effects of nuclear bombs are apparent, measured in kilotons or megatons of conventional explosives, the energy potential of nuclear power plants is also huge. Apart from the source of that power, however, a nuclear power plant uses very conventional means of generating electricity – the heat source turns water to steam which turns a turbine which produces electrical current.

Fukushima Disaster

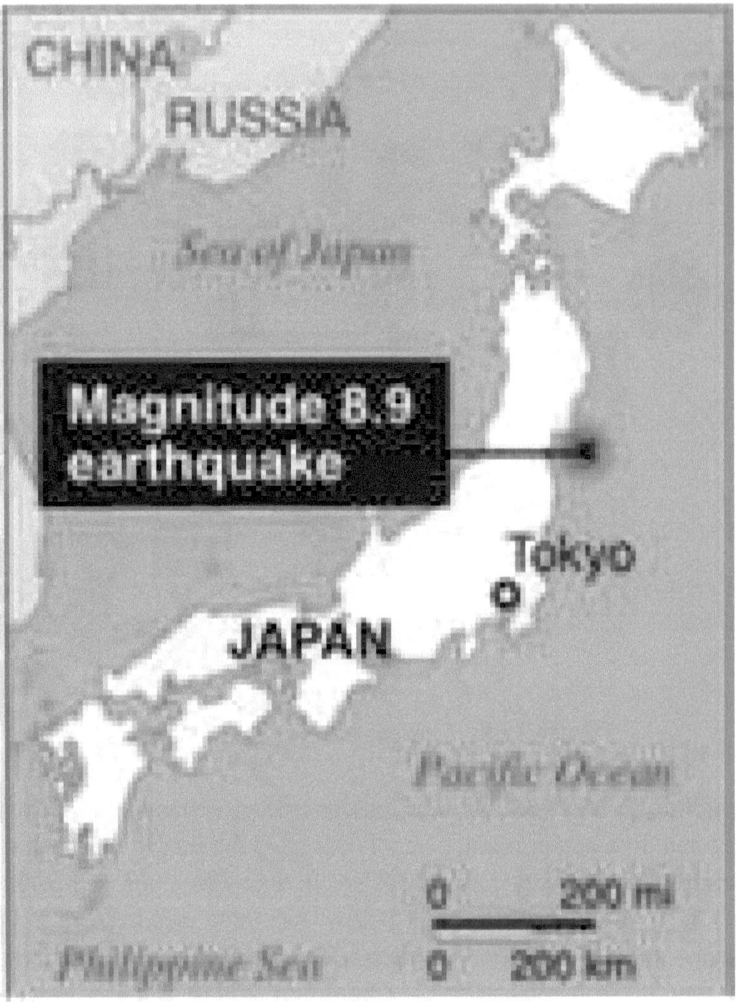

Figure 11.9: ***Earthquake in Japan unleashing Tsunami, 2011.***

On March 11, 2011, a major earthquake hit Japan just off the coast of Honshu, Japan's largest island (www.wikipedia.org/wiki/fukushima_Dalichi_nuclear_disaster). Millions felt it. The nearest town on Honshu was Sendai, 240 km northeast of Tokyo and 130 km from the epicenter of the off-shore quake. The quake was magnitude 9.0, one of the most intense of the last half century. Highways and rail lines, the power grid and buildings were damaged or destroyed. Because earthquakes are common in Japan, the country has strict building codes which averted even greater damage. In Tokyo, buildings swayed, but didn't fall. Even the strictest building codes, though, couldn't prevent what came next, a massive tsunami.

Figure 11.10: ***Whirlpool in ocean after Tsunami & its impact on shore.***

The tsunami swept aside houses, tumbled cars and floated ships well inland, even stranding one large fishing boat atop a two story building. The death toll, mostly from the tsunami, exceeded 20,200 people. Parts of the shore dropped a meter, maybe more, adding to the effective height of the wave. Besides the loss of life and property, a major loss was to the six nuclear reactors at the Fukushima Dai-Ichi (Fukushima Number One) complex. If something goes wrong at a nuclear plant, there are safety features and backups. If something goes wrong with the backups, it's a disaster. That's what happened at Fukushima.

The reactors shut down, promptly and automatically. When sensors registered the temblor, control rods dropped into the reactors, halting the nuclear reactions. Then the tsunami hit, starting a cascade of failures. A protecting sea wall was 6.5 metres high, good enough for the historic quake records of a 1000 years, but inadequate this time; the crest was at least 7 metres, probably topping 10 metres, and the wave swamped the plant.

Even a halted reactor generates heat. Circulating water removes that heat but if the power is out, the water in the reactor boils away. Water also shields radiation; when the circulation is lost, radiation levels rise. Backup generators are supposed to continue pumping cooling water but fail when flooded. Switches and power cables were flooded and inoperable. Battery power is another backup, but batteries run down. Three of the reactors had core meltdowns.

Figure 11.11: ***Fukushima before & after Tsunami.***

The reactors weren't the only problem. Spent fuel rods were stored on site in a pool of water. With all systems failing, additional water wasn't getting to the pools. With lower water levels and higher temperatures in the reactors and the pools, hydrogen was released by reaction of water with zirconium cladding of the fuel rods. The hydrogen gas combined with oxygen and exploded, blowing a hole in the concrete shell of a building housing a reactor. When holes and cracks opened in the containment and pressure vessels of reactors 1, 2 and 3, contaminated water leaked from the structures. To stabilize the water levels after the failure of the backup systems, helicopters flew over, dumping water. The attempt was ineffective and stopped after a few days. Fire trucks were brought in; streams of water from their hoses were directed into the buildings with limited success. Adequate pure water wasn't available. Sea water was used, even though the salts in the water could damage the reactors, possibly clogging the cooling channels and worsening the problem.

Radiation was a concern from unshielded fuel rods. Gases, hydrogen, xenon and others, carry radiation when they escape or are vented from the reactor buildings. Contaminated water had to be stored and containers weren't big enough. Some contaminated water was released to the ocean. Workers had to be monitored and protected from excessive exposure. The helicopters had lead shielding installed to reduce exposure to the flight crews. Estimates are that the Fukushima accident released about 10% as much radiation as Chernobyl.

Because radioactive material escaped, a mandatory evacuation zone was set at 20 kilometres from the complex. Suggested evacuation stretched another 20 km for vulnerable people, especially the young. There were many other effects: with Fukushima shut down, rolling blackouts were needed to conserve power, and sale of vegetables from the region was banned because of radiation worries. Radioactive cesium contaminated tea in the prefectures (states) around Tokyo. Some industries shut down. Even if industries weren't directly affected, some relied on parts manufactured in the damaged area. Across the Pacific, a two-meter tsunami wave struck one town in the United States. Fearing that radiation would arrive next, U.S. citizens near the Pacific coast cleared the shelves of iodide tablets. Concern, about radiation and food safety, was often poorly matched to the actual danger. It will take years, decades even, to bring fully close the plant. Reactor accidents are rated on a scale of zero to seven. Both Fukushima and Chernobyl, but no others, rate a seven. The problems at Fukushima raise concern over the safety of nuclear reactors, not just in Japan, but everywhere.

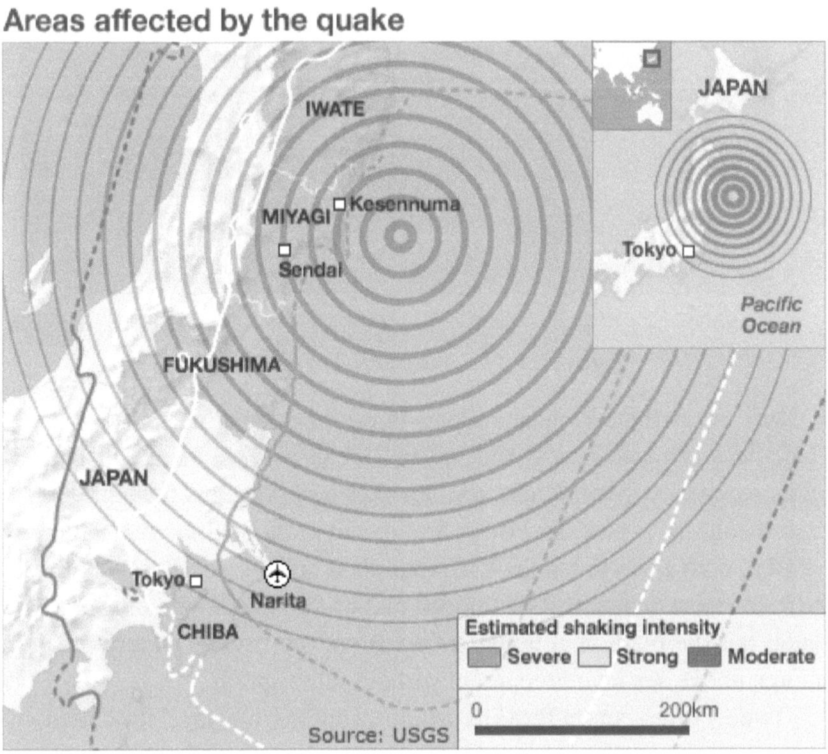

Figure 11.12: ***The Japan earthquake epicenter was about 100 mi. (160 km) east of Sendai, which is about 200 mi. (320 km) NNE of Tokyo.***

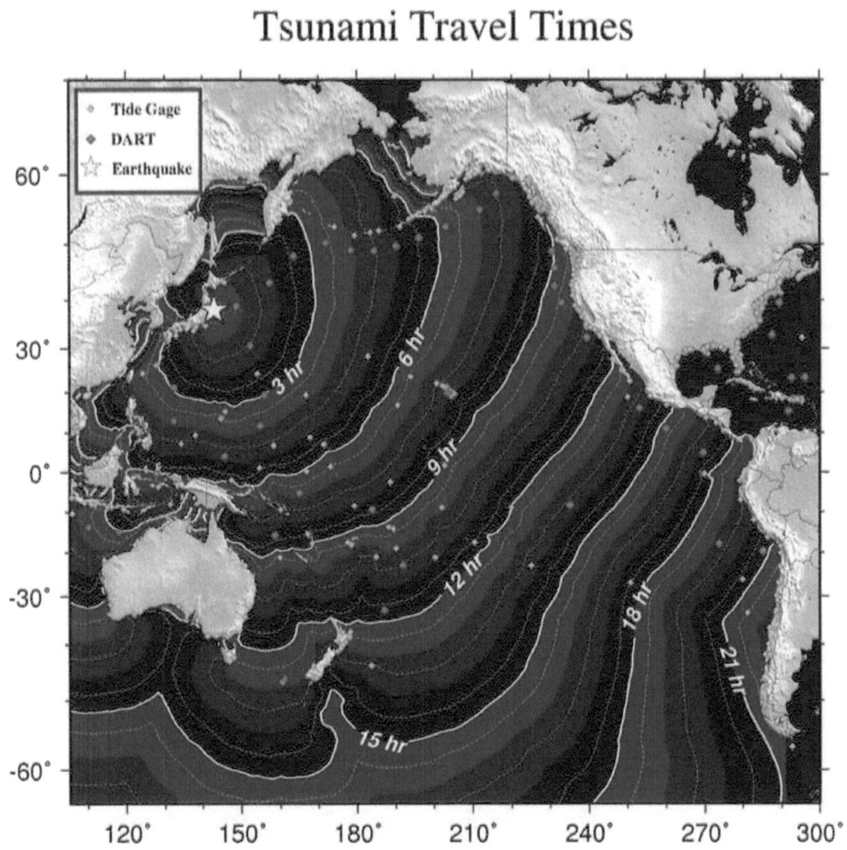

Figure 11.13: ***Tsunami Travel Time.***

In 2004 an earthquake near Indonesia raised a tsunami that also struck Thailand and raced across the Indian Ocean to Sri Lanka and parts of India, especially Tamil Nadu. Earthquakes and tsunamis have previously caused concern at Nuclear Power Stations in India. In 2001 the magnitude 6.9 Bhuj quake was felt at the Kakrapar plant in Gujarat. The tsunami from the Indonesian quake in 2004 flooded part of the Kalpakkam plant in Tamil Nadu, but the plant shut down safely. Also, the backup generators were enough above sea level to operate without interruption.

Twenty-five years before Fukushima, a nuclear reactor at the Chernobyl power station in the Soviet Union had a catastrophic failure (www.wikipedia.org/wiki/chernobyl_disaster). Whhile the two accidents had different causes and results, both happened when multiple breakdowns of design and accident response, each a "perfect storm". Each catastrophe is rated the maximum 7 on the International Nuclear Events Scale. Since the accident, the Soviet Union broke up and the site is in current Ukraine. Pripyat was the town built to house the workers and their families and had to be evacuated as radiation spread from the accident. Pripyat is now a ghost town, unoccupied and decaying, visited only by officials and tourists who take a daylong tour from the Ukraine Capital of Kiev.

As damaging as the two accidents were, neither was as lethal as an atomic bomb. A bomb is designed to release its energy within seconds and it is the blast that does most of the damage to people and property. Those who survive the blast will receive enough radiation to cause illness. While a power plant has more material than a bomb, the arrangement of the fuel means it cannot go off like a bomb. Much of the longer term danger from the two accidents in the radiation induced cancer.

11.3 Oil Spills and Impacts on the Environment

If you drill, beware of the spill, should be the slogan of any company drilling for oil and gas. The spill may happen at the well, in transport, or storage. The spills can be small, a few barrels, or can be huge, a large ship. The oil and shipping companies do not want the spills as any accident costs them money. When they lose a ship, losing the cargo costs them, bad publicity follows, and lawsuits are likely.

While oil floats, wind and waves mix it in a complex fashion. Some oil evaporates, but much of the oily sheen stains whatever it touches or forms scattered tar balls. Pictures of oil coated birds or dead turtles are dramatic, and the volume lost in a major spill is startling. On 28th July, 2003, the Tasman Spirit ran aground near Karachi, Pakistan, and 12,000 tonnes of oil spilled into the Arabian Sea, contaminating 16 km of the coastline. There have been similar oil spills in the Bombay High Coast of Maharashtra. As an even larger spill than the Tasman Spirit and the largest on record for ships is the Atlantic Empress. The ship collided with the Aegean Captain, also a tanker, in July 1979, near Trinidad and Tobago. Around 300,000 tonnes were lost from the two ships; fortunately, none of the oil reached the shore. Much oil is probably in the Empress lying in the deep water.

Accidental spills also range from barely noticeable to astonishing. The worst oil spill in Asia was in Uzbekistan. In March 1992, a well blew out in the Mingbulak oil field and caught fire, burning for two months. Unburnt oil, some two million barrels (which will fill a large oil tanker), was also captured in containment ponds.

In April 2010, near Louisiana in the United States, a well surged out of control while drilling in the deep waters of the Gulf of Mexico. The Deepwater Horizon, a floating drilling platform, exploded and burnt. The fire continued until the rig sank two days later. The sinking destroyed the connection from the wellhead to the surface, and substantial leaking began. As the wellhead was 1500 m down in the ocean, control was difficult. Even knowing what was happening below was difficult.

The amount of crude oil spilled in this incident is probably greater than that from even the biggest tanker accident. Every week, more than 40 tonnes of oil was coming out, the amount carried by a medium size tanker. Oil continued gushing out for twelve weeks. As the oil floated, it covered 6,500 to 24,000 square km, depending on the wind and currents. BP (formerly British Petroleum) was responsible for the operation. They did not deal well with the public perception; at one point, the CEO even said the spill was not that significant. The facts showed otherwise. BP was not sufficiently prepared to anticipate and prevent the accident. A blowout preventer did not stop the flow. The preventer had not been properly tested, and alarm systems were not operational. Once the accident occurred, BP was not prepared to deal with the emergency. The first attempt to stop the leak was to activate the blowout preventer. That failed. Drilling fluid was pumped into the well to plug it, and a large containment dome was installed. Both failed. After 3 months, a second containment cap seemed to halt the flow. Two relief diverted the flow at the cost of $100 million each. The wells would reach 5500 m. The operation used ships, underwater vehicles, helicopters, and air planes at a total expense of several billion dollars. Not just the valuable oil that was lost, but the fisheries that were lost, contaminated beyond use. Beaches were off limits; even hearing of the spill kept tourists away. Many birds and turtles were coated with oil and died.

Petroleum is dangerous when it burns uncontrollably and toxic when it spills. When technology is pushed to the limit (in the Deepwater Horizon case, wells were drilled in deep water), accidents can happen. The results, both anticipated and unexpected, always mean trouble.

Activities and Exercises

1. How has the Chipko movement affected environmental activism in India and around the world? Will you participate in an environmental protection march in your community to improve the air and water quality in your

community? If you will, state the reasons and how the march will impact government policies in your community and the country?

2. What is happening with protection of India's forests, and what laws are used by the Central government to prevent destruction of our forest resources? Write a 2-page report on this topic.
3. How many Tiger Preserves are in India and how are they helping with tourism? Why do you think it is important to preserve wildlife in India? What are the Green Tribunals (Courts) in India effective in protection of tigers and their habitats?
4. How has the Swachh Bharat Mission improved water quality in various parts of India? Research government reports/publications on the accomplishments of the Swachh Bharat Mission. Write a 4-page report on this topic and illustrate your report with graphics and pictures. Are there any such National Missions in India at present to improve the environment and provide the basic needs of all Indian citizens.
5. Research the amount of nuclear energy produced in India, and where are the nuclear power reactors located? Has India had any nuclear power accidents? What are the positives and negative aspects of nuclear energy? Write a 5-page report on the subject and illustrate your report with graphics, charts and pictures.
6. Have a debate in your class about nuclear power? Should it go forward because it eliminates carbon dioxide emissions or be prohibited because of safety concerns?
7. Learn the terminology involved with nuclear power and energy. These include fission, fusion, meltdown, isotope, half-life, and many more. Also, learn the various terms used to measure radioactivity.
8. What are the many uses of radiation and radioactivity, including X-rays and treatment of cancer?
9. Get more details about the Japanese earthquake and tsunami. Make a slideshow or video of the damage. What was the reaction in India? Describe the relief effort, including the role of the government.

   Update the human tragedy and the recovery of the ravaged towns and land.

   Learn the facts about the reactors at the Fukushima complex. Decide what was wrong with the safety features and layout of the reactors. Update the story of the efforts to stabilize the damage at Fukushima. These efforts will be on-going.

What is the reason for using potassium iodide pills when there's a radiation release? How effective are the pills and should the use be limited?

Make sense of the 'alphabet soup' of companies and agencies with involvement in the nuclear industry in Japan, India and the world: JAIF, TEPCO, AEC, AERB, BARC, DAE, NDMA, NDRF, IAEA.

10. About India, make a map of nuclear power reactors - current, under construction and projected. Include as much information as possible: type of reactor, size of reactor (in megawatts), age of the installation, safety concerns. India Today, the March 28, 2011 issue, has a map of locations in an article reporting on the Fukushima disaster. How much of the electricity in India comes from nuclear plants?

    Have the reactors had any accidents, or incidents of not meeting safety standards?

    What are potential hazards for the reactors - tsunamis, earthquakes, aging or poorly maintained facilities? Where are spent fuel rods stored?

11. In case of an accident, is there a disaster plan? What is the evacuation zone? What facilities and equipment are needed and available, including radiation monitoring and protection. What would be the housing and other support for evacuees? How many trained personnel are available and what training do they have or need? Is this a concern where you live?

12. The December, 2004 Indonesia quake killed over 200,000 people, mostly from a tsunami. How much time passed before the wave hit India and Sri Lanka. Where did the wave hit and what was the loss of life and property. What was the effect on the nuclear power plant at Kalpakkam in Tamil Nadu?

13. What parts of India are most likely to be affected by earthquakes, either directly or from a tsunami originating elsewhere. Use http://quakes.globalincidentmap.com/ to see about recent quakes affecting India and the rest of the world. What should India do to prepare for conceivable calamities? What disasters might strike your home area?

14. How much of India's oil and gas needs is fulfilled by domestic production? From where does India import oil and gas to meet its needs? What are the dangers inherent in transporting oil and gas from abroad and within India? Document any oil and gas spills that have

occurred in India? Write a 5-page report on the topic and include graphics and pictures to make the report easily understandable.

References

http://en.wikipedia.org/wiki/Chipko_movement
https://media.springernature.com/original/springer-static/image
http://en.wikipedia.org/wiki/Project_Tiger
http://www.wpsi-india.org/images/tigers_2018s.jpg
http;//en.wikipedia.org/wiki/Swachh_Bharat_Mission
http://wikipedia.org/wiki/fukushima_Dalichi_nuclear_disaster
https://www.theguardian.com/environment/2020/aug/10/mauritius-calls-for-urgent-help-to-prevent-oil-spill-disaster#img-1

Chapter 12

The Future Is Now

Introduction

We can imagine a future that will have minimized carbon emissions to the extent where the global temperature rise will be below 1.5 degree Celsius. That future is now, if we take certain steps to use our current technologies, to re-arrange economic priorities and to bring about new social thinking. In addition, we can learn from nature using biomimicry principles, which is already happening in many fields. The United Nations General Secretary has called the current situation to be an emergency demanding bold action. Dr. Michael Mann's article in the Scientific American magazine (April, 2014) says "Equilibrium Climate Sensitivity(ECS) is a guide to the harmful effects of temperature, which is bound to rise above two °C, if we continue emitting carbon dioxide at the usual pace". Figure 13.1 shows the probabilities of reaching or exceeding two degrees C as assessed by the Intergovernmental Panel on Climate Change in their Assessment Reports of 2007 (AR4) and 2013 (AR5).

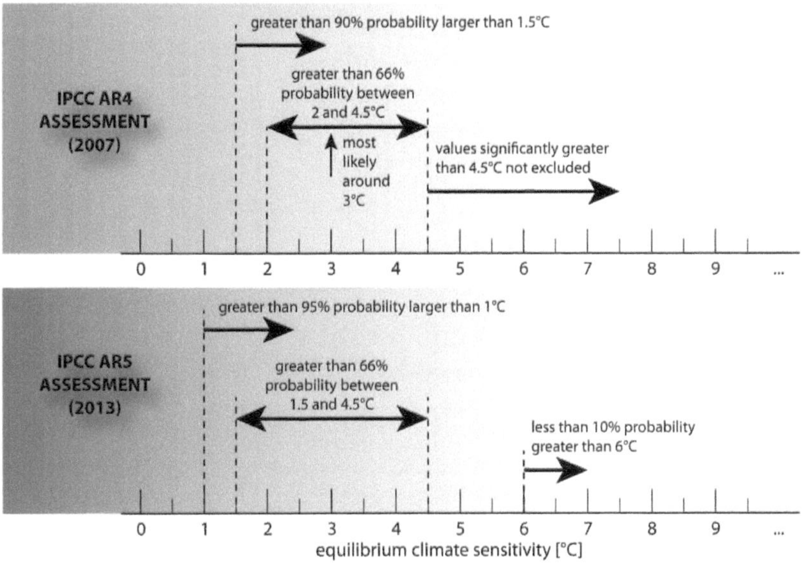

Figure 12.1: ***Graphical representation of the equilibrium climate sensitivity assessment statements of the Fourth (AR4) and Fifth (AR5) assessment reports of the Intergovernmental Panel on Climate Change (IPCC). The AR4 assessed ECS to be likely (greater than 66% probability) in the range between 2 and 4.5°C, very likely (greater than 90% probability) larger than 1.5°C, with a most likely value (mode) around 3°C, while values significantly higher than 4.5°C could not be excluded. AR5 assessed the ECS to be with greater than 66% probability between 1.5 and 4.5°C, with more than 95% probability larger than 1°C, and with less than 10% probability greater than 6°C. Background colors are visual guides only, without any scientific meaning.***

Dr. Mann's chart of the likely, most likely and very likely ECS values for various analytical models is shown in Figure 12.2.

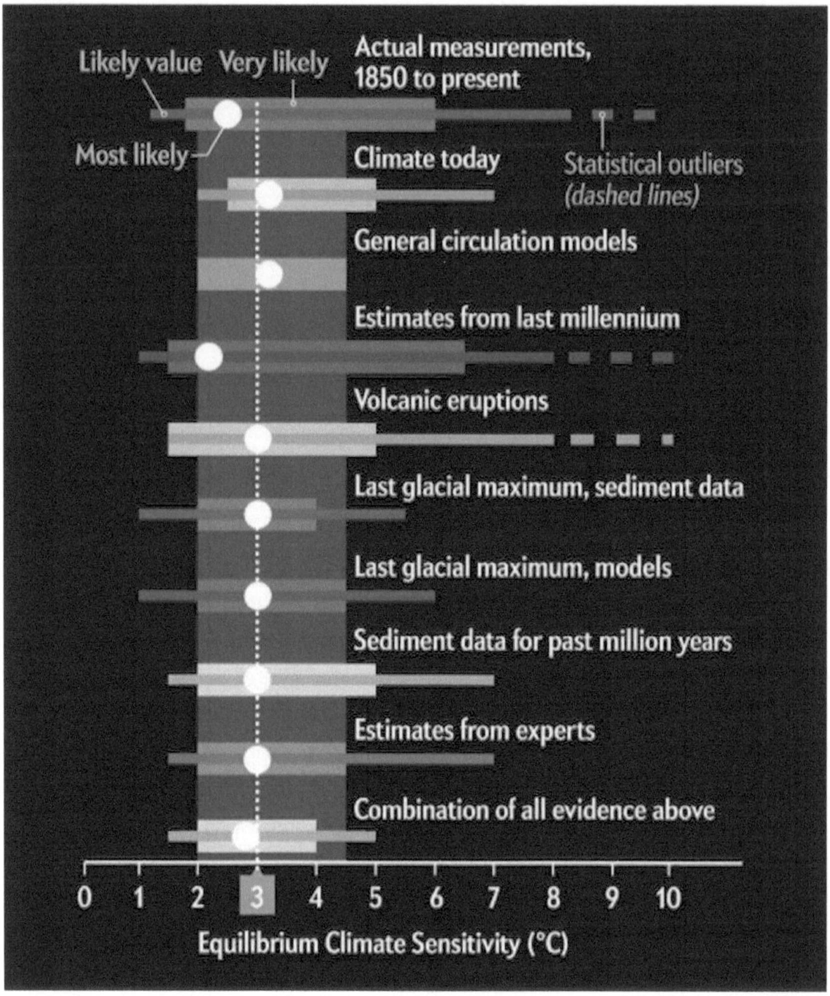

Figure 12.2: ***Equilibrium Climate Sensitivity from various models.***

Some countries are taking bold actions now, and are reversing or slowing the trend of greenhouse gas emissions. A concerted global effort is required to slow the rise in global temperatures so that low-lying areas are not flooded by rising oceans. The technologies for renewable energy and storage of energy should be disseminated widely so that all countries can adopt them and reduce the emissions of greenhouse gases. These issues will be discussed in this Chapter.

12.1 Economic Issues

Adequate economic incentives will change behavior of consumers and generators of energy. A carbon tax has been discussed in the United States for many years, but sufficient consensus has not been reached to make such carbon tax the law of the land. Canada has passed a law attaching a price to carbon and this is promoting the conversion of energy generation from fossil fuels to renewable energy. The Federal Greenhouse Gas Pollution Pricing Act (GHGPPA) was passed in December 2018 and implements a tax which is applied only to provinces and territories whose carbon pricing system does not meet Federal requirements (https://en.wikipedia.org/wiki/Carbon_princing_in_Canada). In provinces where the tax is levied, 90% of the tax revenues are returned to the taxpayers. The other 10% will go to support particularly affected sectors, including small businesses, schools and hospitals. The tax is levied on the carbon content of fuels, and provinces and territories can create their own system of carbon pricing based on needs and requirements of their own jurisdictions.

According to a report by Canadian Chamber of Commerce (CCC) released on December 13, 2018, "the GHGPPA offers flexibility and is the most efficient way to end cut greenhouse gas emissions". The University of Ottawa says that the "CCC's endorsement of the carbon tax will boost Canada's investment in clean technology at home and abroad, and provide the Canadian economy with a major opportunity to market itself in a low carbon future". The price of carbon begins at $20 per tonne in 2019, and will rise to $50 per tonne of carbon dioxide by 2022 (https://en.wikipedia.org/wiki/Carbon_pricing_in_Canada). Carbon pricing in Canada is expected to remove 50 to 60 MT of emissions from the air annually by 2022, which represents about 12% of all Canadian emissions. However, Canada needs to reduce emissions to 512 MT by 2030 to meet its Paris Climate Change commitment. This means an annual reduction of 200 MT from the 2018 levels, and so, the government is pursuing a range of other policies, including improving fuel standards, energy efficiency, and closing coal-fired plants.

British Columbia was the first to introduce a carbon tax in 2008. Since its introduction, fossil fuel use in the province fell by 16% compared to a 3% increase in the rest of Canada, and its economy outperformed the rest of Canada. This has proved that the carbon tax benefits were "no longer theoretical" and that they did not hinder economic growth. As more countries follow Canada's example, it will have a significant impact on the accumulation of greenhouse gases in the atmosphere. As more countries such as India and China promote the use of renewable energy and energy efficiency, the trend is for falling greenhouse gas emissions and the reduction of the current peak of 415 ppm of greenhouse gas concentration in the atmosphere. Businesses have an incentive to reduce greenhouse gases, according to an article in the Sierra magazine of May/June 2019 (Sierra magazine, Vol. 104, No. 3, 2019). It states that extreme weather events and natural disasters caused $91 billion in damages in 2018 alone in the United States, and the latest National Climate Assessment warns that if the world does not take strong action soon, annual losses to US businesses could total $500 billion. A United Kingdom nonprofit, called Carbon Disclosure Project, has collected climate risk data from nearly 7000 companies globally (1800 are in the US). In 2017, 70% of the companies in the Standard & Poor's 500 companies disclosed climate risks their companies face. This will result in investors demanding action to manage climate change and this will result in further reduction of greenhouse gas emissions.

An article entitled Banking for Justice in the Sierra magazine (Sierra magazine, Vol. 104, No. 3, 2019) describes the verdict in jam vs. International Finance Corporation (IFC), in which the US Supreme Court said that "international organizations like the IFC do not enjoy virtually absolute immunity and therefore can be sued. Athialy, a lawyer for the New Delhi-based Center for Financial Accountability, states that "the judgment will strengthen communities' efforts to hold the bank accountable and is a step in the direction of bringing accountability in financial institutions". The IFC and other international banking institutions are wary of funding fossil fuel based power plants and are funding renewable energy projects. Thus, economic issues are driving our response to climate change and this trend will accelerate as data from more extreme weather events and adverse impacts on poor communities is available.

David Wallace-Wells (2019) states in his book "The Uninhabitable Earth" that unchecked wisdom of the market has to be replaced with moral values of international cooperation to save the planet for future generations. The Paris Climate Agreement provides a framework for international cooperation to reduce greenhouse gas emissions, and transfer of funds from the developed countries to developing countries that do not have all the resources needed to transition the economy to a carbon-neutral economy. Developed countries should take this promise seriously to help the developing countries and live up to their promises made at the Paris Agreement. In addition, cooperation is needed on technology transfer so that green and renewable energy technologies are adopted worldwide.

New trade deals between nations should stress the moral infrastructure of climate change, by encouraging the reduction of carbon footprint of goods traded through economic incentives. Ideological disputes among nations should be replaced with consensus driven action toward minimizing greenhouse gases. Intellectual property rights related to technologies for green and renewable energy, increased food production, and water management should have special treatment so that the technology transfer becomes easier and new technologies are adopted faster in countries around the world. The quality of life and reduction of economic inequality should take precedence over the constantly growing Gross Domestic Product of a country.

12.2 Policy Issues

The Paris Agreement has prompted many governments to adopt bold policies to reduce the greenhouse gases. India, China, France, Denmark, Great Britain and Germany have aggressive policies in place to promote renewable energy and reduce the use of fossil fuels. The US government is lagging behind; however, the State of California, many US cities and private companies are taking aggressive steps to reduce greenhouse gas emissions. The State of California has signed an agreement with 4 major auto companies that improve the gas mileage of cars and trucks (for reducing greenhouse gas emissions) although the US government is trying to relax the gas mileage standards nationally. The Danish government is working closely with Indian government and companies to achieve the following:

- Identify opportunities for technology transfer to India for offshore wind energy and biomass in India
- Expand wastewater treatment and water supply demonstration projects in India, including the use of sewage sludge
- Strengthen the marketing of energy and environment technologies using the public-private consortium State of Green (stateofgreen.com)
- Work with Tata companies, India's largest business group, in the areas of water consumption, purification of wastewater, sustainable construction, wind energy and energy efficiency

- Work with Suzlon, India's largest and the world's fifth largest wind turbine manufacturer, to improve wind turbine technologies.

Such technological and economic cooperation is required between developed countries and developing countries around the world to minimize greenhouse gases, increase the use of renewable energy, and improve efficiency of energy and water use.

Governments around the world are passing laws and introducing policies which reduce the use of fossil fuels, provide economic incentives for promoting solar and wind energy, and promote use of rainwater harvesting and water conservation. These policies have to be enforced through public education and monitoring so that they can make a positive impact on the availability of water and energy to poor communities that are unduly impacted by climate change.

Individual consumer choices that promote reduction of the carbon footprint cannot be a substitute for political action by governments around the world. Politics of cooperation instead of division should be demanded by citizens around the world. Citizens have to become politically active and demand policies that promote quality of life through clean air, water and land. Participation in civic society organizations, volunteering to help the poor in a community, and actively participating in voter registration and voting in local, State and Central Government elections should be promoted by bold policies adopted by government agencies. In democracies, elections can change the course of the government and citizens should elect candidates who will reduce income inequality, and provide incentives and funding for green infrastructure and renewable energy initiatives.

12.3 Technology Issues

Technology is advancing rapidly and many centers of excellence for renewable energy, water, and Artificial Intelligence (AI) are doing great work around the world. The use of organic materials to convert them into biodiesel or biogas is progressing well in the US, India, China and Europe. If we can convert any kind of biomass (instead of just corn) into ethanol and other biofuels, it will transform the energy industry. Much of agricultural waste is either burned or just left on the ground and if this could be mobilized for producing liquid or gaseous biofuels, it will alleviate poverty in rural areas while enhancing the overall availability of energy for transportation and other needs. Solar technologies are being developed at an accelerated pace, and the cost of solar panels is coming down, making solar competitive with other forms of conventional energy. The use of wind farms on land and in the ocean is progressing well, and technology for improving energy efficiency of wind turbines and storing the produced energy is being developed.

The use of AI is growing rapidly and software companies are spending a lot of money to develop new applications for AI. Improvements in the following areas, made possible by AI, can make a significant impact on climate change efforts and improve the quality of life:

1. By monitoring our energy and water usage on a daily basis, we can live comfortably without wasting energy or water. We can immediately notice changes and make the necessary improvement to prevent water leakage, conserve water and reduce our energy usage. This can be done in individual homes, public and private buildings, museums, and many other recreational facilities.
2. By monitoring our daily traffic patterns in cities, we can minimize traffic gridlock and thus save a lot of energy in the transportation sector. This will increase productivity since people will not be stressed by sitting in traffic, and reach their destination by the fastest safe route.
3. We can improve operations of our water and wastewater facilities, electric utilities, manufacturing plants, agricultural practices, and all aspects of life.
4. We can improve healthcare and provide access to the poorest of the population so that prevention can become the norm instead of the expensive treatments the current healthcare system is advocating.
5. We can monitor the air quality and amount of carbon dioxide and other greenhouse gases in the atmosphere and regulate the emissions to keep under the target of 415 parts per million and maintain the temperature increase to under 1.5 degrees C.
6. Improvements in agricultural practices that minimize water and energy use, and enhance productivity of the land can be achieved by using AI technologies.

New technologies should be transmitted to users worldwide through international cooperation and use of the internet. Many universities in the United States and private groups have put a lot of educational materials on the internet to enhance the quality of education among citizens around the world. Similarly, the technologies developed for improved energy and water management, and green infrastructure should be made available through the internet. Political cooperation and selfless thinking by private corporations can enable this transformation. For example, Khan Academy.org has put a lot of valuable information for Kindergarten through 12th grade education on the worldwide web.

12.4 Social Issues

Behavioral change is a slow process and as local communities get involved in the management of climate change impacts, behavioral change among citizens is possible. It all starts with education of our young generation in schools and colleges. Student activism is increasing around the world and if this activism is continued later in their lives, our increase in greenhouse gases can be curbed and the temperature rise kept under 1.5°C. The increased use of solar and other renewable energy in communities can put market pressure on industry to reduce greenhouse gases. Each person in the community should have a stake in the welfare of the community and things will drastically change. Environmental activism has been responsible for the improvement of the quality of life in cities around the world. Such techniques should be used to change the behavior of individuals, local governments, state governments, Federal or Central governments and finally international institutions like the World Bank, United Nations, etc.

Educational awareness campaigns are necessary for the people to understand the urgency of Climate Change and its impact on their lives. Such campaigns against smoking, unsafe cars, and health issues have modified people's behavior and the same has to be done for Climate Change. The Climate Reality Project (www.climaterealityproject.org) in the US has conducted several educational campaigns around the world and it is gaining momentum in communities and state and national governments are responding to the message of taking action to minimize climate change impacts. Many other organizations are also working toward this goal.

Activities and Exercises

1. Study the National Climate Action plan published by the Indian government and write a summary report on how India plans to reduce greenhouse gases and improve the environment.

2. Imagine a world where technologies for renewable energy production, distribution and transmission are available to all. Write a paper on how this will reduce the extremes of weather and minimize the impacts of climate change in your city.
3. What actions will you promote in your classroom and in your school to reduce the carbon footprint? How will you monitor this and what result do you expect from these actions?
4. What is being done in your city to monitor the air pollution and minimize the health impacts related to poor air quality? If you were given an opportunity to reduce air pollution in your city, what actions will you recommend?
5. What economic incentives could be provided by the local government to encourage recycling of all solid waste generated in your city? What action can be taken by individual households to recycle all solid waste that they generate?
6. Research the Artificial Intelligence technologies that are being used in your community to improve traffic movement, and reduce air and water pollution. Do you think we can achieve a driverless car in the future? What problems do you anticipate in using driverless cars in your city?
7. How has the internet changed our lives? Imagine a future where the internet is used to improve the quality of life in your city. Write a report with your findings.
8. How can individual actions in a community reduce the level of poverty and improve income inequality? What can the government do to help the homeless and provide basic services such as education, healthcare and food security?
9. What can every citizen in your community do to reduce water use, conserve energy and recycle solid waste? How will you start a tree planting program in your community to improve air quality, and make your community green? What are the impediments to starting such a program?
10. What can we learn from the wisdom of the people in our rural communities who have been living in harmony with nature for many years? Research www.honeybee.org. Write a short report on how simple technologies mentioned in the website can be used in your community.

References

- Wallace-Wells, David, 2019, The Uninhabitable Earth: Life after Warming, Crown Publishing Group, a division of Penguin Random House LLC, New York.
- Mann, Michael, 2014, Equilibrium Climate Sensitivity, Scientific American magazine, April.
- Sierra Magazine, 2019, Vol. 104, No. 3, May/June.
- https://en.wikipedia.org/wiki/carbon_pricing_in_Canada.
- Figure 13.1 Courtesy: Intergovernmental Panel on Climate Change (IPCC)
- Figure 13.2 Courtesy: Michael E. Mann (michaelmann.net › content › sensitive-topic) https://www.scientificamerican.com/article/mann-why-global-warming-will-cross-a-dangerous-threshold-in-2036/

Chapter 13

Epilogue

Even as we conclude this book, more innovations and breakthroughs are happening with Energy and Environmental Technologies which are reducing greenhouse gas emissions around the world. As this book goes to print, we will have a new President in the United States of America (US) Mr. Joe Biden who believes strongly in Climate Change and Renewable Energy Technologies. He will join the Paris Climate Change Treaty which the US decided to pull out of during the previous administration. Mr. Biden has promised to reduce the greenhouse gas emissions to net zero by 2035. China and India are taking leading roles in implementing renewable energy technologies and energy efficiency. European nations are aggressively working toward net zero emissions in their economic planning. Exciting battery technologies for efficient storage of solar energy and biomethanation technologies for converting various organic wastes into biofuels are in the works at many universities and industrial companies. Airlines are converting to more planes with energy efficient engines that can use biofuels.

Indigenous and traditional lifestyles across the globe have always modeled reverence for the environment and their wisdom is profoundly clear to all. The future is bright for reducing greenhouse gas emissions since young people have realized the impact of global warming on their futures. As was mentioned earlier in the book, a young 16-year old girl from Sweden, Greta Thunberg, has raised awareness of global warming and its impacts around the world and in her speech at the United Nations, expressed the hope that all nations will work toward a greener future. Energy efficient technologies are being spread through the internet to all parts of the globe and many local leaders are taking steps to transition to non-fossil fuels. Just in a few years, the economics of the coal industry have changed dramatically as power plants using coal as a fuel are being shut down and no new ones are being built. International lending institutions are careful in approving development projects around the world which do not adequately evaluate the impact of the project on greenhouse gas emission.

Our book has highlighted new technologies and policies, and social institutions that are focused on reducing the carbon footprint of our activities. The Climate Reality Project which was started by Mr. Al Gore has chapters around the world, educating the school children and adults about the global warming impacts, and what actions can be taken by the government and citizens to reduce the greenhouse gas emissions. If concerted effort can be taken by governments around the world and by global corporations to reduce greenhouse gas emissions, and promote energy efficiency and renewable energy, we can foresee dramatic improvements in our environment by 2050. Such action will ensure that we meet the targets of the Paris agreement and may be, start on new agreements which will further reduce the impacts of climate change. The destruction to forests and property, and loss of human and animal lives during large wildfires and storm events has pushed us further into immediate actions which can save our planet from further global warming.

Our book has highlighted many technologies such as Artificial Intelligence and Renewable Energy, and these will continue to be further improved through human innovation. The cooperation among different countries and companies to solve the problem is increasing every year. Advances in telecommunications such as the 5G internet connectivity and wide-spread use of smart cell phones is making instant communication possible among various organizations, even during the Covid pandemic which has put at an end to personal meetings this year. The United Nations is taking an active role in the dissemination of knowledge regarding Climate Change and bringing together scientists through the International Panel on Climate Change. Our hope is such efforts will further expand in the coming years, and we might be hopeful of a day when we have achieved or exceeded our targets for global temperature increase, and start healing the planet.

It is our sincere hope that this book will be widely used in India and in other countries to raise the awareness of Climate Change among high school students, who are the future of the planet. We have made it easy for teachers to challenge their students for innovative thinking in this area by providing Activities and Exercises at the end of each chapter. Hands-on learning will create the interest in knowing about our environment and what humans are doing to the environment. It is hoped that the students using this book will become activists for a greener planet and urge local, state and Central Governments to take action to stop the increase and maybe reduce greenhouse gas emissions over the coming years. As future consumers, they will exert pressure on manufacturers and global corporations to reduce their carbon footprint.

With high hopes for a brighter future where the global impacts of Climate Change are minimized and further adoption of green technologies will take place worldwide, we close this book.

Index

4. Energy Demand

5. Biodiversity Protection

6. Ecosystems

7. Floods

8. Environmental Crisis

9. Habitat Conservation

10. Environmental Approaches

www.ingramcontent.com/pod-product-compliance
Ingram Content Group UK Ltd.
Pitfield, Milton Keynes, MK11 3LW, UK
UKHW041637031026
15587UKWH00001B/338

9 798905 609244